W9-AVT-217

R.F. DIGGAN
JOHN DEERE
FARM IMPLEMENTS

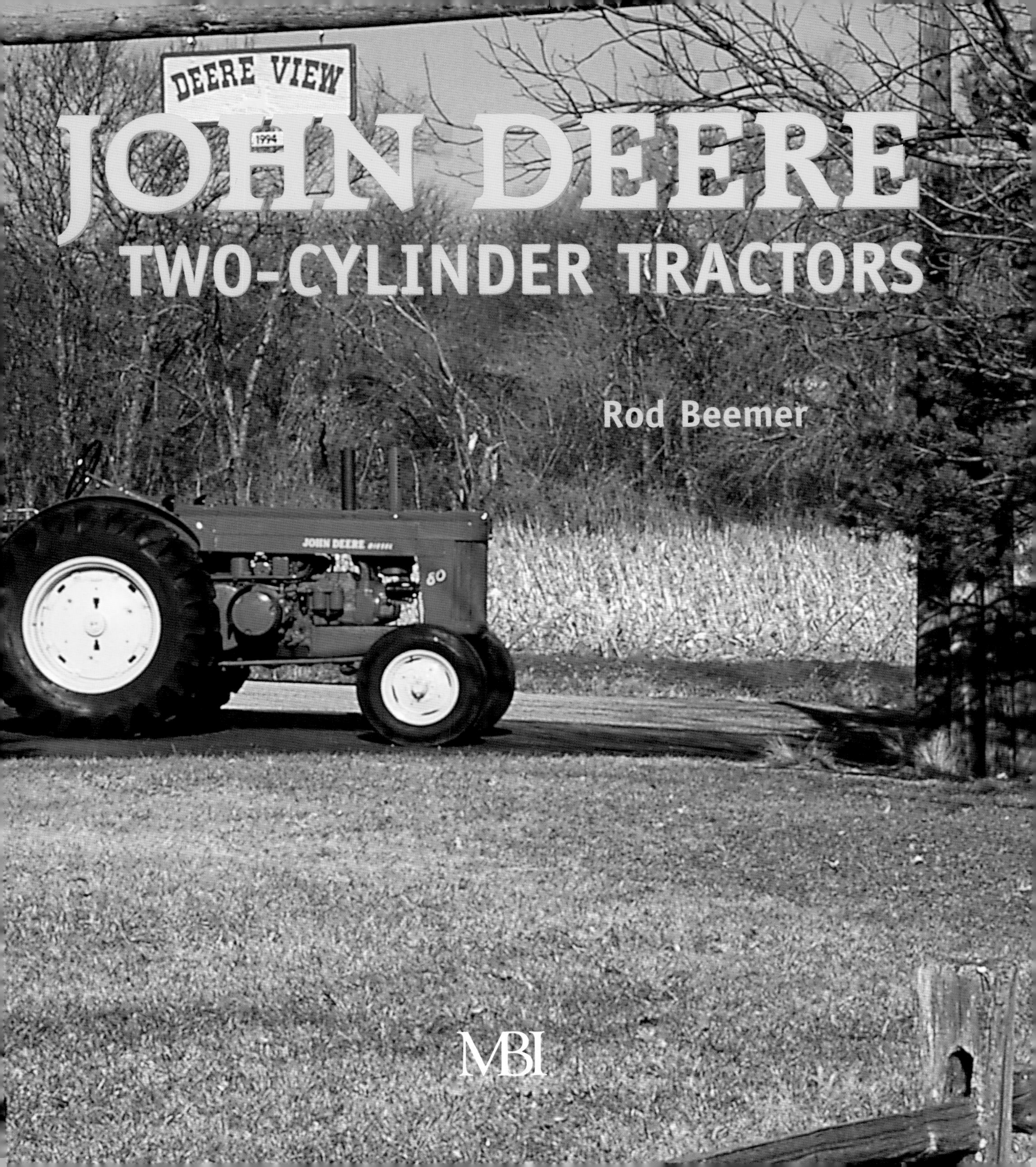
DEERE VIEW
1994
JOHN DEERE
TWO-CYLINDER TRACTORS
Rod Beemer
JOHN DEERE
80
MBI

## *Dedication*

To Leonard

This edition first published in 2003 by
MBI Publishing Company, Galtier Plaza, Suite 200,
380 Jackson Street, St. Paul, MN 55101-3885 USA



ISBN 0-7603-1619-8

**On the front cover:** This 1956 Model 70 Standard Diesel is equipped with PTO, three-point hitch, adjustable wide front axle, and a V-4 gasoline-starting engine.

**On the frontispiece:** Several credible collectors have acknowledged that this is a very rare John Deere dealership sign. The present owner speculates that it's from the 1930s or early 1940s. Originally, it graced a John Deere dealership located just outside of Philadelphia, Pennsylvania.

**On the title page:** A 1956 Model 80 Diesel.

**Table of contents page:** A Model MT weighs in at about 2,600 pounds, stands 56 inches high at the radiator, and is 110 inches long. Sure, it isn't a big tractor, but it looks like a toy next to a 27,000-bushel grain bin.

**On the back cover:** A very nice 1936 Model A on wide-rim, round-spoke F&H rear wheels. The owner traded two Model MT tractors for this completely restored Model A.

Edited by Amy Glaser

Printed in China

# Contents

# Acknowledgments

A grateful thanks go to the following collectors and restorers: Mel Kopf; Kenny Read; Justin Read; Charles Dugan; Albert Jansen; Bob Roblin; Jake Delaney; Clifford Smith; Lester, Kenny, and Harland Layher; Duaine and Orville Filsinger; John Nikodym; LeRay Koons; Ron Jungmeyer; Mrs. Kenneth Peterman; Dan Peterman; Don Nolde; Earl and Harold Hartzog; Leon and Theresa Beedy; Mark Strasser; Leo Zeigler; Andy and Karen Anderson; and Larry Shetter.

Thanks also to Mike Mack, retired director, Product Engineering Center at Deere & Company; and Bill Bulow, retired, Deere & Company Waterloo Foundry.

These dedicated people help keep the history alive.

# Introduction

They are all in this book, from the Waterloo Boy to the 30 Series models, but this isn't a tech manual on John Deere two-cylinder tractors. Technical information on these tractors is readily available from numerous other sources. Rather, this is a book about how these tractors fit into the larger drama of a company, a nation, and a world that underwent tremendous change between 1918 and 1960.

## One of America's Oldest

John Deere two-cylinder tractors' roots run deep within the history of the United States. When John Deere was born in 1804, there were only 17 states in the Union, and agriculture was still a primitive industry that required prodigious amounts of human and animal muscle to put food on the nation's tables.

At the same time William and Sarah Deere welcomed their son John into this world, Meriwether Lewis and William Clark began their epic Journey of Discovery. Their mandate was to explore President Thomas Jefferson's remarkable Louisiana Purchase and evaluate its potential for the nation.

The Louisiana Purchase more than doubled the size of the United States by adding 827,192 square miles of real estate. The prairies and plains included in this vast expanse would eventually become the breadbasket of the United States and much of the world.

Years later, Deere & Company's two-cylinder tractors provided part of the mechanical muscle necessary to transform these prairies and plains into productive farms. However, a lot of company, national, and international events would unfold before John Deere tractors became a reality.

In 1815, when John Deere was 11 years old, Napoleon Bonaparte met defeat at Waterloo, Belgium. One-hundred and three years later, Deere & Company would become linked with its own Waterloo—Waterloo, Iowa. Unlike Napoleon's Waterloo, which ended in defeat, Deere & Company's Waterloo began a remarkable story of success.

John Deere probably wasn't even aware of the death of England's King George III in 1820, because at the time, John Deere was beginning his apprenticeship to the blacksmith trade. Nine years later, in 1829, Andrew Jackson, known as "Old Hickory," took office as the seventh president of the United States. That same year, Deere opened his first blacksmith shop, in Leicester, Vermont.

Davy Crockett, Jim Bowie, and 187 other men were immortalized at the Battle of the Alamo in 1836,

# Contents

# Acknowledgments

A grateful thanks go to the following collectors and restorers: Mel Kopf; Kenny Read; Justin Read; Charles Dugan; Albert Jansen; Bob Roblin; Jake Delaney; Clifford Smith; Lester, Kenny, and Harland Layher; Duaine and Orville Filsinger; John Nikodym; LeRay Koons; Ron Jungmeyer; Mrs. Kenneth Peterman; Dan Peterman; Don Nolde; Earl and Harold Hartzog; Leon and Theresa Beedy; Mark Strasser; Leo Zeigler; Andy and Karen Anderson; and Larry Shetter.

Thanks also to Mike Mack, retired director, Product Engineering Center at Deere & Company; and Bill Bulow, retired, Deere & Company Waterloo Foundry.

These dedicated people help keep the history alive.

# Introduction

They are all in this book, from the Waterloo Boy to the 30 Series models, but this isn't a tech manual on John Deere two-cylinder tractors. Technical information on these tractors is readily available from numerous other sources. Rather, this is a book about how these tractors fit into the larger drama of a company, a nation, and a world that underwent tremendous change between 1918 and 1960.

## One of America's Oldest

John Deere two-cylinder tractors' roots run deep within the history of the United States. When John Deere was born in 1804, there were only 17 states in the Union, and agriculture was still a primitive industry that required prodigious amounts of human and animal muscle to put food on the nation's tables.

At the same time William and Sarah Deere welcomed their son John into this world, Meriwether Lewis and William Clark began their epic Journey of Discovery. Their mandate was to explore President Thomas Jefferson's remarkable Louisiana Purchase and evaluate its potential for the nation.

The Louisiana Purchase more than doubled the size of the United States by adding 827,192 square miles of real estate. The prairies and plains included in this vast expanse would eventually become the breadbasket of the United States and much of the world.

Years later, Deere & Company's two-cylinder tractors provided part of the mechanical muscle necessary to transform these prairies and plains into productive farms. However, a lot of company, national, and international events would unfold before John Deere tractors became a reality.

In 1815, when John Deere was 11 years old, Napoleon Bonaparte met defeat at Waterloo, Belgium. One-hundred and three years later, Deere & Company would become linked with its own Waterloo—Waterloo, Iowa. Unlike Napoleon's Waterloo, which ended in defeat, Deere & Company's Waterloo began a remarkable story of success.

John Deere probably wasn't even aware of the death of England's King George III in 1820, because at the time, John Deere was beginning his apprenticeship to the blacksmith trade. Nine years later, in 1829, Andrew Jackson, known as "Old Hickory," took office as the seventh president of the United States. That same year, Deere opened his first blacksmith shop, in Leicester, Vermont.

Davy Crockett, Jim Bowie, and 187 other men were immortalized at the Battle of the Alamo in 1836, the same year that John Deere was fighting his own personal battle—for financial survival. He lost that battle and left his home and family in Vermont, and, like thousands of other hopeful pioneers, he headed west to start anew.

Deere settled in Grand Detour, Illinois, and within a year had opened another blacksmith shop, where he built his first steel plow that would scour the region's prairie soil.

As the saying goes, the rest is history.

While Deere labored to turn iron and steel into plows, the nation labored with slavery; the Trail of Tears; the burning of Lawrence, Kansas; the Civil War; the assassination of President Lincoln; the California gold rush; the Chisholm Trail; Custer's Last Stand; and the addition of 21 states to the Union.

There is good evidence that Deere & Company is the second-oldest continuously operating manufacturing firm in the United States. Without question, it is the only full-line U.S. agricultural implement and equipment-manufacturing corporation that hasn't been merged or sold.

It's doubtful that John Deere, and the others who helped build Deere & Company realized that they were making history. But, viewed from the twenty-first century, that's exactly what they did. John Deere two-cylinder tractors are a major chapter in that history. Consider all the food they helped put on our people's tables, the roadways they helped build, and the building sites they labored on, and it becomes apparent that Deere & Company has touched the life of almost all Americans since those first days.

The next time you drive through that annoying construction on our freeways, you'll no doubt see John Deere equipment at work. Drive by almost any residential or commercial construction site, and chances are John Deere equipment is on the job. As you travel through rural America, it's a good bet that green and yellow John Deere tractors and equipment will be visible in the fields and farmyards. The ancestry of today's ultra-modern, high-tech John Deere tractors goes directly back to the two-cylinder tractors discussed in this book.

During the 42 years of the two-cylinder era, Deere & Company's tractor program went from reactive to proactive. During the first decade, it was mostly "management by crisis." During the last decade of that era, management focused—and succeeded—on becoming number one in the business—and succeeded. A large part of that success was due to its two-cylinder tractors.

While this isn't a tech manual, it is impossible to give even a thumbnail sketch of these tractors without mentioning some numbers. Occasionally, there are discrepancies among sources concerning the actual number of tractors produced for any one model. Production numbers given in this work are approximate only and are not meant to be considered as gospel.

Deere & Company operated much like the automobile industry by introducing a new model in the fall of each year. Like the automobile, the model year of a John Deere tractor is considered to be for the following calendar year. Unless mentioned as an introduction date for a tractor, the dates are model years.

Many changes and improvements were made during the lifetime of the Model R and Model N, but the logo design remained the same. However, the color of aftermarket decals seems to differ, depending on the source. Experts consider the decal with the yellow hue to be more correct.

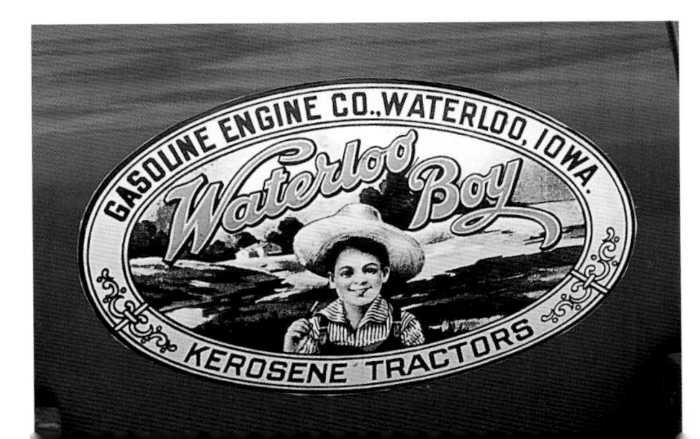

This 1920 Waterloo Boy appears very similar to the Model R that it replaced. This particular tractor is one of the last chain-steer vehicles before the final switch was made to automotive-type steering.

# 1 Old Century, New Century

## Deere's $2.35 Million Dollar Tractor

John Deere died in 1886, and his son Charles stepped to the helm of what had grown into a corporation called Deere & Company.

In 1892, the first quasi-successful kerosene/gasoline-powered tractors appeared on the agricultural landscape. These experimental tractors proved to be a historic landmark in agricultural history. However, it wasn't the only event of the 1890s that affected agriculture and the settling of the West.

In 1892, 1.8 million acres of Montana land belonging to the Crow Indians were opened for white settlement, and 30,000 homesteaders rushed onto three million acres of Indian land in Oklahoma. It all added up to vast new markets for farm machinery manufacturers and, eventually, tractor customers. However, during the last decade of the 1800s and the first ten years of the 1900s, tractors didn't register a very strong blip on Deere & Company's radar screen. The plow, not tractors, was the product that had built the company.

Not unlike the 1990s, the 1890s were a time of business consolidations, expansions, and mergers that gave rise to giant business conglomerates dealing in oil, sugar, whiskey, and a host of other products. The prevailing corporate boardroom axiom was "get bigger or get out." With business under government scrutiny due to the Sherman Anti-Trust Act, the legal form of organization for the empire builders became the holding company. Foreign money bloated the domestic capital reserves, launching round after round of buyouts, mergers, and takeovers of companies engaged in everything from flour to steel.

In 1889, Deere was targeted by just such a group of English investors. They hoped to form a plow trust, or implement manufacturing super concern, by buying and consolidating three major implement companies: Deere & Company, Deere & Mansur, and the Moline Plow Company.

It is interesting that Charles Deere favored the proposal and worked to keep the deal alive through years of on-again, off-again negotiations. Perhaps memories of his father's financial struggles explain Charles Deere's willingness to join forces with other manufacturers. In 1892, though, the deal fell through for the final time.

Deere & Company was caught in another wave of consolidation fever in 1899 when it was one of 20 manufacturers optioned to form the proposed American Plow Company. Once again, Charles Deere favored the proposal, but by 1901, this plow trust went the way of the previous effort. A year later, another attempt was made to organize a trust, but it failed and sounded the final death knell of any formation of a plow trust super concern. Deere & Company would face the future alone, a future that proved to be an unsure and rapidly changing one.

This Waterloo Boy Model R was built on December 18, 1917, and is considered a 1918 model. It was originally shipped to Saskatchewan, Canada. The Model R was introduced in 1913 and replaced by the Model N in 1918.

The biggest single decade for legal immigrants entering the United States, with 8,795,386 new arrivals, was between the years of 1901 and 1910. The vast majority, just over eight million, were from Europe, according to the *1994 Statistical Yearbook of the Immigration and Naturalization Service.*

These new Americans wanted to own land—an impossible dream in their native countries. Consequently, many became soldiers in the trans-Mississippi's vast army of homesteaders. All were potential new customers for Deere & Company and its competitors.

Implement manufacturing underwent major changes in the years prior to this, with the number of firms decreasing from 2,116 in 1860 to 715 in 1900. By 1910, the figure had dropped to just 640 active agriculture implement manufacturers, which meant fewer companies to provide more product for an escalating demand.

In 1902, the McCormick Harvesting Machine Company, the Deering Harvester Company, and the Milwaukee Harvester Company merged to become the International Harvester Company (IHC). By 1903, this new firm controlled about 90 percent of the grain binder production and 80 percent of the mower production in the United States.

Charles Deere died in 1907, at which time the company was a small but respected player in the farm implement industry. The board of directors elected William Butterworth as company president. This year was also one of deteriorating economic conditions for the nation. Unemployment was high, banks failed, and food prices rose dramatically. However, the "Panic of 1907" didn't last long, and Deere had sufficient cash reserves to weather the storm and still show a profit.

Butterworth and the board of directors faced a business climate much different than founder John Deere or his son Charles had faced. The industrial revolution had spawned giant corporations—in farm machinery, IHC dominated the industry. In the auto industry, it was Henry Ford, who launched the famous Model T in 1908 and was on his way to a high place among the giants. A decade later, Ford threw his considerable wealth and talent into tractor manufacturing. His Fordson was responsible for plowing under many farm implement-manufacturing firms during the 1920s.

In 1909, IHC's assets of $172.7 million placed it fourth in size among all U.S. corporations. Only U.S. Steel, Standard Oil, and the American Tobacco Company listed more assets. Of course, IHC was right in the middle of the growing tractor mania. The Davids of the farm implement industry, Deere included, were now face-to-face with Goliath, IHC.

Charles Deere realized that Deere & Company must grow or be swallowed by the competition, and he was willing to explore merger possibilities. In July 1909, Butterworth and Deere's board of directors followed the same path by approaching J.I. Case Threshing Machine Company with a proposal for consolidation. It failed.

The next opportunity for consolidation and company security came from none other than IHC, with

This view of the Waterloo Boy Model R shows the chain-steering arrangement. All Model Rs used chain steering, but not the Model N.

These glass jars located above each rod bearing offer a visual check to show the operator that the oil pump is working. They are a marvelous example of off-the-shelf technology—¼-pint cream bottles that were common in the era of glass milk bottles and horse-drawn dairy delivery wagons.

plows for harvesters the pitch. IHC didn't manufacture plows, and Deere didn't manufacture harvesting equipment. Each needed the other's product in its sale catalog to be a full-line company. However, this effort at cooperation and consolidation also didn't yield any fruit.

Was it fate, or was it just a string of fortuitous circumstances? It was probably some of both. It was also probably some luck mixed with the conservative bent of Deere's board of directors. Not to be dismissed is Butterworth's zealous regard for Deere's family business. For whatever the reasons, merger wasn't in the cards for Deere, and history proved it a good thing.

In 1910, tractors were not a very big part of the company. Nevertheless, in January 1910, as competition mounted and consolidation efforts faltered, the board of directors decided to bring all Deere interests under complete control of the parent company. It also resolved to purchase the required manufacturing firms necessary to make Deere a full-line company and therefore control its own destiny. These decisions had a big impact on the company's future and on the coming tractor issue, and they proved to be defining decisions by Deere management.

The implication for tractors? Had Deere been content to become a short-line manufacturer—making primarily plows and tillage implements—and if it managed to survive, there's little historical evidence that any John Deere tractors would have ever been built. Only full-line companies were successful in the tractor-manufacturing arena.

Consequently, the board's decision in 1910 precipitated a round of acquisitions that brought 10 companies into the Deere fold during 1910 and 1911.

The company had purchased two manufacturing firms previous to this decision. In 1907, Deere & Company purchased the Fort Smith Wagon Company of Fort Smith, Arkansas. This acquisition was followed in 1908 by the purchase of the Marseilles Manufacturing Company of Marseilles, Illinois. As the name implies, the former manufactured wagons, and the latter made a line of automatic corn shellers and a line of grain elevators.

Joseph Dain pioneered a line of hay handling equipment that was considered the best in the business. His Ottumwa, Iowa, company, Dain Manufacturing Company, was purchased by Deere & Company in 1910.

The manufacturing firms that Deere purchased in 1911 included the Moline Wagon Company, Van Brunt Manufacturing Company, Union Malleable Iron Company, Syracuse Chilled Plow Company, Deere & Mansur Company, Kemp & Burpee Manufacturing Company, Reliance Buggy Company, and the Davenport Wagon Company.

The new acquisitions were adjusting to being part of Deere & Company when the U.S. Congress passed the 16th amendment to the U.S. Constitution. A federal tax on income now had to be programmed into the financial equation. However, even the hated income tax couldn't overshadow the tractor dilemma that continued to appear on the agenda in Deere's boardroom. It was an issue that wouldn't go away and couldn't be ignored any longer.

Deere responded with a cautious research and development (R&D) tractor program that yielded the Melvin tractor in 1912 and the Dain and Sklovsky tractors in 1915. Not everyone at Deere & Company was excited about, or comfortable with, the prospect of seeing mechanical horsepower replace horses and mules as the power source for John Deere implements.

In fact, Deere & Company President Butterworth bowed his head and resisted any plans for Deere's involvement with the tractor business. In September 1916, he wrote the following instruction to board member George Peek: "I want it plainly understood that I am and will remain opposed to our taking up the manufacture of tractors and will take steps to stop it if an attempt is made to start. . . . If it comes up I want you to stop it." Butterworth also expressed his feelings that 1916 was the time to reduce Deere's product line rather than expand into new or additional products.

Regardless of Butterworth's personal feelings, it was becoming impossible to ignore the accelerating growth of the tractor industry during these years. Consider this from R. B. Gray's work *The Agricultural Tractor: 1855–1950.* "In 1917, the year the United States entered World War I, tractor production more than doubled. A total of 62,742 were manufactured as 85 new companies entered the field to bring the number of manufacturers to more than 200. Approximately 15,000 tractors were exported."

Tractor production more than doubled again in 1918, with 132,000 vehicles produced. It became painfully clear that Deere's competition was selling a lot of tractors and consequently stealing a lot of implement sales. It was decision time at Deere & Company.

Frank Silloway, head of Deere's Marketing Division, suggested to the board that buying an existing tractor manufacturing firm might be preferable to Deere & Company tooling up to build its own tractors. He also knew of a tractor company that was for sale—the Waterloo Gasoline Engine Company, of Waterloo, Iowa. Immediately, the board authorized Silloway to thoroughly investigate this possibility.

To help win the war overseas, President Woodrow Wilson asked for, and received, a draft, which eventually swelled the 200,000-man army to nearly five million. Suddenly, thousands of American farm boys were a lot more concerned about staying alive than they were about tractors and fieldwork.

With so many young men serving in the military, a serious manpower deficit developed in the domestic workplace. Women, 1.4 million of them, stepped forward to shoulder the workload and forever changed the character of this nation. These women were the unsung heroes who kept farms and industry productive during this difficult time. England faced a real and urgent threat of food shortages, and U.S. tractors stepped in to help provide food for the good guys fighting in Europe. The demand for food pushed U.S. wheat production to one billion bushels in 1915.

## Deere's $2.35 Million Tractor

In 1893, John Froelich organized the Waterloo Gasoline Traction Engine Company, of Waterloo, Iowa. Before this landmark, but ill-fated venture, Froelich had operated a threshing business in Iowa and South Dakota, where he used a Case steam-traction engine to power the separator, or threshing machine. His experiments to replace the steam powerplant with a gasoline engine resulted in his first tractor. This new innovation provided the power source for Froelich's threshing machines during the 1892 threshing season. The Froelich tractor performed well enough to attract several local Waterloo investors, and in 1893, Froelich and his partners formed the Waterloo Gasoline Traction Engine Company. However, only four copies of the Froelich tractor were made, and unfortunately they

# Timeline: 1900–1929

**1900**
U.S. census records a population of 75.9 million.

There are over 8,000 automobiles on American roads.

The most deadly U.S. natural disaster occurs when a hurricane hits Galveston, Texas, killing an estimated 10,000 to 12,000 people.

**1901**
President William McKinley is shot by an assassin and dies eight days later. Theodore Roosevelt becomes president.

**1902**
International Harvester Company (IHC) is launched.

**1903**
Orville and Wilbur Wright record first airplane flight.

Henry Ford sells his first Model A automobile.

National Guard is sent to Wyoming to halt the violence between cattlemen and sheepmen.

**1904**
Henry Ford sets speed record of 91.37 miles per hour.

President Teddy Roosevelt is elected to a full term.

The Kinkaid Act opens desert land in Nebraska for homesteaders.

**1905**
The Wright brothers make a 24-mile flight.

The Homestead Act opens 1,500,000 acres of the Wind River Reservation for settlement.

**1906**
Upton Sinclair publishes *The Jungle*, where the meatpacking industry is examined and found rotten.

**1907**
Charles Deere dies.

Oklahoma becomes the 46th state.

J. Pierpont Morgan supplies $25 million to save banks from closing.

**1908**
Country Life Commission is created. It is a federal project to evaluate the quality of rural life—expected findings are that life on the farm is good and getting better.

Ford Motor Company introduces the Model T.

Durant's General Motors is Ford Motor Company's newest competition.

**1909**
700,000 acres of land in Washington, Montana, and Idaho are opened for settlement.

**1910**
Harry Ferguson builds and flies Ireland's first successful airplane. Ferguson's involvement with agriculture impacts Deere & Company on several occasions in the future.

U.S. census places the population at 91.9 million—about one-third, or 32 million, live and work on farms.

**1911**
The U.S. Supreme Court decides that Rockefeller's Standard Oil Company of New Jersey violated the antitrust law. The ruling orders the company to downsize by selling almost 40 companies.

**1912**
Deere & Company's R&D Melvin tractor is built.

**1913**
The 16th Amendment to the U.S. Constitution, "to lay and collect taxes on income from whatever source derived," becomes law.

One of the worst floods in U.S. history occurs along the Ohio River, killing hundreds and leaving thousands homeless.

**1914**
Waterloo Boy Model R is introduced.

World War I begins. President Wilson promises U.S. will remain neutral.

**1915**
Sklovsky and Dain R&D tractors are built at Deere & Company.

Ford assembly line coughs up Henry's one-millionth automobile.

**1916**
Waterloo Boy Model N is introduced.

Foreign trade shoots upward to $8 billion.

**1917**
United States enters World War I.

**1918**
Deere & Company purchases the Waterloo Gasoline Engine Company, builder of the Waterloo Boy tractor.

War rationing includes meat, wheat, and sugar.

**1919**
Allied and Central Powers sign the Versailles Treaty.

Former president Theodore Roosevelt dies.

didn't live up to expectations. Disappointed, Froelich left the company in 1895, and afterward it was reorganized as the Waterloo Gasoline Engine Company.

As the name change suggests, the tractor, or traction, part of the business took a backseat to other products. Stationary gasoline-kerosene engines became the bread and butter products of the new company. Other items in the company's line included cream separators, milking machines, and manure spreaders.

Fortunately, the Waterloo Boy wasn't entirely forgotten by management, and experimental models were kept on the back burner. In 1911, the company's tractor development program got a significant boost when A. B. Parkhurst joined the team. Not only did the company get Parkhurst, it also acquired the three tractors that he had designed and built utilizing two-cylinder, two-cycle horizontally opposed powerplants.

This new engine design was adapted for use on the Waterloo Boy Model L and Model LA. The Model L was a three-wheel vehicle, while the Model LA featured a four-wheel design. Only nine Model Ls and 20 Model LAs were built. The unit's horizontally opposed two-cylinder engine featured a 5.5x7-inch bore and stroke that displaced 333 cubic inches. Maximum rpm was pegged at 750, and the tractor created about 7 drawbar and 15 belt horsepower. Unfortunately, this engine didn't satisfy the customers or the engineers.

The Waterloo Engine Company solved its problem by designing a side-by-side version of the two-cylinder engine. This new design was installed in an LA chassis and introduced as the Waterloo Boy Model R.

As historic events were unfolding on a global scale, Frank Silloway traveled to Iowa and investigated the Waterloo Gasoline Engine Company. Arriving in Waterloo, Iowa, Silloway found a modern plant producing two models of the Waterloo Boy tractor. The Model R sold for $985, while the Model N was listed at $1,150.

This was higher than the board's 1914 directive that a Deere tractor must be built to retail in the $700 range, but the best tractor Deere's R&D program had to offer carried a much higher price tag. Things looked promising for a John Deere tractor if Butterworth could be convinced that Deere & Company needed to get aboard the tractor bandwagon.

The belt pulley is on the left side of the Waterloo Boy, located on the outboard side of the flywheel. On the Model D, which replaced the Waterloo Boy, it was moved to the right side of the tractor and eliminated the hand crank. The flywheel on the Model D and other John Deere tractors became the "crank."

Wayne Broehl's book *John Deere's Company* says that Silloway reported to the board, "Here we have an opportunity to, overnight, step into practically first place in the tractor business. . . . I believe that we would be acting wisely if we purchased this plant."

Driven off the crank, one belt drives the fan and the other belt drives the water pump on the Waterloo Boy Model R.

Despite the fact that the Dain tractor was now in the production stage and despite Butterworth's expressed reluctance, in March 1918, Deere & Company's board of directors voted unanimously to purchase the Waterloo Gasoline Engine Company for $2.35 million. Frank Silloway's strong recommendation carried the vote.

Deere & Company had at last made a commitment to compete in the burgeoning tractor business. For the next 42 years, John Deere two-cylinder tractors were a major influence in the farm machinery industry.

Prior to Deere's purchase, the Waterloo Gasoline Engine Company had purchased the Waterloo Foundry. As a result, Deere & Company now had a production facility, a foundry, and a tractor—a two-cylinder tractor that already had an interesting and successful history.

## Waterloo Boy Model R

Finally, here was the two-cylinder tractor that became the granddaddy of the "Johnny Popper." The Model R went through several style changes during its lifetime. The engine was one of the features that was changed and then changed again.

From 1915 to 1917, the bore was increased to 6 inches, but retained the same 7-inch stroke that produced 396 cubic inches of displacement. Thereafter, the engine had a 6.5x7-inch bore and stroke that upped the displacement to 465 cubic inches.

Just over a year into Deere & Company's new tractor venture, on July 31, 1919, employees walked out of the Waterloo facility in a labor strike, which lasted until December 14 of that same year. Total sales for the fiscal year ending October 21, 1919, were $5.1 million with a net profit of about $200,000, down considerably from the previous year.

## Waterloo Boy Model N

With the introduction of the Model N, the bore and stroke of the engine remained at 6.5x7. Approximately 20,000 copies of the Model N were produced before it was replaced with the John Deere Model D in 1924.

In 1920, a Waterloo Boy Model N had the distinction of being the first tractor tested at the newly established Nebraska tractor test laboratory, where it posted 15 drawbar and 25 belt horsepower.

The above information on the Waterloo Boy tractors is only a general overview. Like all tractor manufacturers of the period, the Waterloo Gasoline Engine Company searched for the best design that would deliver the best performance and sales. Within the model designations, there were also style changes—some of which have few, if any, details.

Total production of about 30,700 Waterloo Boy tractors can be verified, but not all of those stayed within the continental United States. Harry Ferguson's Belfast, Ireland, auto dealership, Harry Ferguson Ltd., began to sell the Waterloo Boy tractors under the name Overtime. Eventually, Ferguson's connection with agriculture would have an enormous impact on the tractor industry.

The Overtime paint scheme was a departure from the standard Waterloo Boy colors. Overtime tractor's fenders and chassis were a similar green, but the

A quick identification feature to distinguish between the Model R and the Model N is the rear-wheel drive gear. This is a Model R, but on the Model N, the gear is directly under the wheel rim.

wheels were red/orange and the engine, fuel tank, and radiator were gray.

Although the Waterloo Boy tractors were manufactured by the Waterloo Gasoline Engine Company, these tractors all burned kerosene, with the exception of the U.K. Overtime, which was powered by paraffin, the British version of kerosene.

For Deere & Company, the years 1910 to 1919 helped define and provide direction for the company. It had explored the idea of consolidation to survive and replaced it with a round of acquisitions that gave it size enough to be a major player. Deere was now a full-line company with a successful tractor in its dealership's showrooms.

This 1924, Model D 26-inch Spoker was originally shipped to Ohio. Three-quarters of a century later, it still looks and runs good. It was purchased and brought to the Great Plains by a Kansas collector.

# 2

# 1920s

## Deere Begins with D

The census of 1920 pegged the U. S. population at 105 million with the farm sector falling below 30 percent. This meant fewer farmers were working larger farms that required larger tractors for both drawbar and row-crop work.

The Waterloo Boy Model R introduced in 1913 underwent around 12 different versions or styles, from the RA through the RN, with the RN becoming simply the Model N. Nevertheless, by the early 1920s, the Waterloo Boy still looked like a Waterloo Boy—an old-fashioned tractor when compared to the competition.

Besides getting a turnkey tractor operation when Deere & Company purchased the Waterloo Gasoline Engine Company, Deere also discovered a very valuable bonus—a new model undergoing R&D at the Waterloo plant. Deere personnel hadn't been told this before the sale, and the surprise turned out to be extremely fortuitous for the new owners. This experimental tractor eventually became the famous John Deere Model D.

### The Mechanical "Animal"

It took another five years before the new model was deemed ready for production. By this time, the postwar economy was extremely depressed, and Deere's tractor sales had fallen dramatically in the past two years.

Deere had sold only 79 tractors in 1921 and 307 tractors in 1922. Consequently, projections for tractor sales during 1923 and 1924 were revised even further downward. This caused some board members to question the wisdom of building the Model D during such a depressed economy.

Despite this negative environment, board member Leon Clausen believed in the future and pushed for a production run of at least 1,000 Model Ds. Perhaps Clausen recognized the huge appetite that Americans were displaying for the mechanical "animal." Fifteen million automobiles were roaring up and down the nation's roads by 1920, and the United States was producing two-thirds of the world's oil. His persuasive arguments finally convinced the board it was the right thing to do. Consequently, Clausen should be remembered as a key figure in Deere's tractor program and a person instrumental in launching the Model D.

However, when Warren Harding took the presidential office in 1921, the nation's economic woes were worse than Deere's. On top of the country's financial crisis, a prominent political figure and future president, Franklin D. Roosevelt, fell victim to a disease that was terrifying the nation—polio.

The good news was that Deere & Company, the economy, and Roosevelt would all enjoy brighter days ahead.

It weighs about 185 pounds, measures 26 inches across, has 6 open spokes, and spins at 800 rpm. It looks great captured on film, but the spoked flywheel demands respect when it's running and you're standing close by.

## The Model D Does It

Spoker D; nickel-hole flywheel; corn borer special; two-, three-, or four-hole spoke steering wheel; ladder-side radiators; and streeters are all part of the descriptive and colorful collector jargon that has emerged around the venerable John Deere Model D.

This tractor, introduced in 1923, established Deere & Company as a serious contender in the tractor business. Although it underwent numerous upgrades through the years, it was a simple and durable tractor that stayed in the John Deere line for 30 years.

Just two months after Model D production began, President Harding died, on August 3, 1923. Upon Harding's death, Vice-President Calvin Coolidge became the nation's 30th president.

In 1923, Congress passed the Agricultural Credits Act, which created 12 Federal Intermediate Credit Banks alongside the 12 banks of the Federal Reserve System. Each bank was funded with $5 million to lend to farm cooperatives who could then loan the money to farmers. Some of these dollars no doubt found their way into Deere's coffers as payment on a Model D tractor.

The unstyled Model D (1924–1938) was a bare-bones, bare-knuckled, no-nonsense, simple workhorse that tested the mettle of any operator who spent hour after hour and day after day on its pitching, rolling, spartan deck. Its short wheelbase and low stance were mindful of a pugnacious bulldog.

By rotating the engine 180 degrees from the Waterloo Boy's placement, the Model D's crank was positioned at the rear of the engine, next to the transmission. This allowed the belt pulley and the flywheel to be mounted directly on the crankshaft—the belt pulley on the tractor's right and the flywheel on the left. This change also placed the cylinders and spark plugs toward the front of the tractor.

Most Model D veterans would probably agree that there's a great deal of truth in an old farmer's advice of, "If you can find a one-owner John Deere D, buy it, 'cause there ain't no single man that can wear one out." They could take a lot of punishment and keep on popping.

In many aspects, a Model D is a D—all 159,083 copies. This is an approximate number, and about 100 were Industrial versions. Of course, there were changes and improvements throughout its long production run.

The Model D has the distinction of having the largest cubic-inch displacement of any two-cylinder John Deere tractor, at 501 cubic inches from a 6.75x7-inch bore and stroke that was used from late in 1927 through the end of production. Prior to that, the horizontal two-cylinder was a 6.5x7-inch bore and stroke providing 465 cubic inches of displacement.

Model D tractors burned almost any fuel, including kerosene, distillate, or tractor fuel. Toward the end of production, it could be optioned into a gasoline burner at the factory, or the necessary parts could be ordered for a dealer conversion. The Model D debuted with a 2/1 transmission, which remained unchanged until 1935, when it was replaced with a 3/1 design.

The first 50 copies of the Model D had a fabricated front axle that quickly proved inadequate for the rigors of fieldwork. Beginning with the 51st tractor manufactured, the fabricated axle was replaced with a cast iron axle.

In 1924, at the same time the 26-inch flywheel was replaced with the 24-inch flywheel, a new transmission case was adopted that allowed a power takeoff (PTO) to be offered as a dealer-installed option and later as a factory-installed option.

Striving to keep pace with the industry, Deere built 97 improved experimental tractors in 1928. These XD tractors have an "X" preceding the serial number on the serial number plate. One of the major upgraded features on these units was the three-speed tranny. Another round of 50 experimental Model Ds was built in 1930, and the serial number plates on these rare tractors have a "B" prefix to the serial number. At this time, the engine rpm was bumped from 800 up to 900.

In 1932, the Model D had a brief flirtation with crawler tracks when the Lindeman Power Equipment Company, of Yakima, Washington, fitted three Model D tractors with crawler undercarriages. The Best Manufacturing Company, of San Leandro, California, manufactured these undercarriages. The Best firm was one of the two California pioneers in manufacturing track-laying vehicles and eventually merged with the famous Caterpillar Tractor Company, of Peoria, Illinois.

Beginning in 1933, rubber tires became available, which was another step taken by the engineering department to keep the flagship of the John Deere line competitive.

Industrial applications for the Model D were recognized soon after production began, but it wasn't until 1935 that an official Model DI was recognized. Optional industrial equipment, including solid rubber tires and high-speed final drive sprockets and chains, were available as early as 1926.

The official DI models built between 1935 and 1941 were painted industrial yellow with black silk-screened letters. The Model DI was never styled.

Three unstyled Model Ds were tested at Nebraska. Test number 102 in 1924 recorded 22 drawbar and

The spark plugs and priming petcocks are almost identical to the Waterloo Boy's, as seen on this 1924 Model D. The Dixie-Arrow magneto is also very similar to the Waterloo Boy's. The air cleaner is a Donaldson.

30 belt horsepower using kerosene as fuel. Test number 146, conducted in 1927, produced 28 drawbar and 36 belt horsepower using kerosene. The last Nebraska test on the unstyled Model D was test number 236, performed in 1935. This time, the fuel was distillate and the horsepower 30 at the drawbar and 41 on the belt.

In 1939, this previously simple, functional workhorse hit the nation's farms sporting a bit of style. The Model D, following closely on the heels of the Model A and Model B, became the third John Deere tractor to be styled by Henry Dreyfuss. New tinwork, including fenders and dust shields, made the operator's platform much cleaner, and a new instrument panel made reading gauges and reaching controls much easier. The most welcome optional equipment features were electric starting and an improved lighting package.

At the introduction of the styled Model D, rolling on steel front and rear was standard equipment, although rubber tires front and rear were an option, as was a combination of steel and rubber. Rubber tires became standard equipment in 1941, and the rear-tire size was changed from 28 to 30 inches. The years during World War II brought another wheel and tire change—back to steel as standard equipment—due to the war effort's priority on the rubber supply.

Regular production of the Model D ended in March 1953 after a stellar 30-year run. Several factors contributed to its demise, but the introduction of the Model R was probably one of the biggest internal considerations. It was easier to bring a completely new model onboard rather than redesign the Model D to meet the market demands of more horsepower, better transmissions, and improved hydraulics.

An early Model D with the decal on the end of the fuel tank. The decal specified the correct lubrication procedures. Farmers were just learning about mechanical power and the necessity of proper maintenance. By placing the decal on the tank, the operator had many hours to contemplate the important instructions.

This 1925 Model D hasn't yet received the decal for the end of the fuel tank. However, it does show the coveted Waterloo Boy transmission cover.

Today, at least one restored Model D can be seen among the tractors of most serious John Deere collectors. Old-timers always have a bit of nostalgia in their voice when they discuss the various rare features found on the early Model Ds.

## Spoker D

The heavy, exposed cast flywheel on the left side of green and yellow tractors is vintage John Deere, and they came in several sizes and styles on the Model D. The first 1,279 vehicles came with a 26-inch, 6-spoke open design, which was followed in 1924 by a 24-inch version of the spoked flywheel. Approximately 4,968 copies of the 24-inch style were manufactured until late in December 1925, when the spoked flywheel was replaced with a solid design. A few of these spoked flywheel models have survived and are referred to as Spoker Ds.

### *Nickel-Hole Flywheel*

Immediately following the spoker design came the nickel-hole or small-hole flywheels. The term solid "flywheel" technically is not correct, because a completely solid wheel would crack as the cast iron cured or was subjected to heat, cold, and operating stress. To alleviate this problem, a stress slot with a hole at each end was cast into the flywheel. The hole at each end of this slot in the small-hole flywheel measured ⅞-inch, or about the diameter of a nickel, hence the term nickel-hole flywheel. Time and experience proved that many, if not most, of the approximately 2,036 flywheels cracked. Increasing the hole diameter

## Timeline: 1920–1929

**1920**
Some autos found on the showroom floor are Chalmers-Franklin, Maxwell, Mercedes, Milburn Electric, Packard, Peerless, Pierce-Arrow, Salient, Stephens, and Stutz, along with some old standbys like Ford, Chevrolet, and Cadillac.

**1921**
Franklin Delano Roosevelt contracts polio.

Canadian biochemists find treatment for diabetes insulin from a pig's pancreas.

**1922**
Oil fields in Wyoming are subject of Teapot Dome scandal.

**1923**
First two-cylinder John Deere tractor is introduced—The Model D.

Agricultural Credits Act is passed, making loans available to farmers.

U.S. has 15 million registered automobiles.

U. S. Steel begins eight-hour workday.

**1924**
John Deere Model D: 1924–1953. Waterloo Boy tractor is discontinued.

**1925**
Forty percent of nation's workers earn more than $2,000 per year.

Henry Ford's assembly lines turn out an automobile every 10 seconds.

Walter Chrysler launches the Chrysler Company.

**1926**
One in six Americans owns an automobile.

**1927**
Pilot Charles Lindbergh crosses the Atlantic.

Henry Ford pulls plug on "T" and introduces the Model A. The Model T racked up an amazing 15 million copies.

Ford discontinues U.S. production of his Fordson tractor.

**1928**
John Deere Model C: 1928. John Deere Model GP: 1928–1935.
Automakers Chrysler and Dodge merge.

Herbert C. Hoover is elected president with promise of "a chicken in every pot and a car in every garage."

John Deere Model GPWT: 1929–1933. General Motors purchases German automaker Opel AG.

Ford's Model A is a success, with one million copies off the assembly line.

International Harvester stock soars to $142 per share, and the company posts record profits of $37 million dollars. Stock market crashes.

at the end of the stress slot to 2 inches eliminated the problem of cracking.

Bill Bulow, retired employee of the Waterloo foundry, said that the old flywheels were probably straight cast iron, and cast iron has 0 percent flexibility, so it required very little heating and cooling to stress the cast iron enough to cause a crack.

### *Ladder-Side Radiator*

The first 50 Model Ds had four distinctive squares in the radiator side frame, which gave the appearance of a ladder with rungs. Another important change was the fabricated front axle was replaced with a cast-iron version. These 50 tractors had serial numbers 30401 through 30450.

Three slots are visible on the steering wheel spokes of this 1924 Model D. Upon introduction of the Model D in 1923, the steering wheel had four slots.

This 1925 Model D steering wheel has only two slots in the spokes, and for the following model year, the slots had disappeared completely.

### *Four-, Three-, and Two-Spoke Steering Wheels*

Another feature on the first 50 Model Ds was a steering wheel with four spokes, with each spoke having four elongated holes in it. Each subsequent year saw the number of holes reduced, from four to three, three to two, and, finally, solid spokes.

### *Corn Borer Special*

When the destructive corn borer struck the Corn Belt in 1927, the U.S. Department of Agriculture Bureau of Entomology purchased 444 Model D tractors to help power the department's campaign aimed at controlling the pest. Since the government program mated these tractors to a McCormick-Deering Stubble Beater, they were all equipped with a PTO shaft.

### *Streeters*

Soon after Model D production closed in 1953, Deere learned that some farmers were still insistent on purchasing a Model D tractor. Unfortunately, the assembly line was being dismantled, but enough parts were found in the plant for 92 more vehicles. These last-of-the-last Model Ds were assembled by hand in the street between the mill room and truck shop, hence the name Streeters.

Nebraska test number 350 conducted in 1940 was the last official test of the Model D. Burning distillate fuel, the results were 38 drawbar and 42 belt horsepower.

## Deere's First General Purpose Tractors

Tractors were beginning to bedevil farm implement manufacturers just as William Butterworth assumed the presidency of Deere & Company in 1907. Old, full-line companies like Deere & Company and International Harvester (IH) had built empires on horse-drawn and horse-powered equipment. The words "John Deere" and "plow" were almost synonymous due to the product that launched the company—the steel plow. By 1909, Deere was producing approximately 1.25 million plows per year—all horse drawn.

But, times were changing. By 1918, Deere & Company reluctantly conceded to mechanical power. Henceforth, the Waterloo Boy, the Model D, and future models were destined to pull John Deere plows.

Plows weren't the only implement that farmers wanted to hitch behind this newfangled power source. Predictably, row-crop farmers were soon looking for a suitable tractor to pull planters and cultivators. In 1909, Deere produced more than 900,000 cultivators, and not one functioned properly behind a Waterloo Boy or a Model D tractor.

However, just a year after the Model D was introduced, IHC showcased its new Farmall, which revolutionized row-crop farming. It was painfully obvious that Deere needed a row-crop tractor to guarantee that John Deere implement customers wouldn't defect to the competition. Deere also recognized that the Waterloo Boy had outlived its usefulness and discontinued its production in 1924. This left only the venerable Model D in Deere's tractor stable.

Although some effort was made to adapt the Model D to row-crop work, the Model D was first and foremost a drawbar tractor designed as the power source for pull-type implements or as power for belt applications.

Developmental work on John Deere's All-Crop tractor began in 1925, which, like the Farmall, was aimed at doing all the various jobs on a farm, especially planting and cultivating row-crop.

## Model C

Early in 1927, the All-Crop name was changed to the Model C, and production began in March 1928. It took only until April to build the approximately 99 tractors that bear the Model C designation. The company recalled at least 53 of these to be rebuilt and renumbered.

The Model C flathead engine was a 5.75x6-inch bore and stroke displacing 312 cubic inches and rated at 950 rpm. It was a kerosene burner with water injection into the carburetor to help minimize detonation under heavy load. A 3/1 transmission provided forward and reverse travel. Highly touted in company advertisements were the four power applications provided by the Model C: drawbar, PTO, mechanical power lift, and belt pulley.

Drop housings on the rear axle elevated the Model C's rear axle enough to accommodate row-crop. Since the Model C Standard was a three-row outfit, an arched front axle allowed clearance for the row that was straddled. There were some doubts about the three-row design, and some experimental versions of the Model C were produced with a tricycle front axle and designed to accommodate two- and four-row implements.

## Model GP

The new model designation GP (General Purpose) was introduced in August 1928 and reflected the general purpose role of the tractor. Necessary mechanical changes were made to correct any problems that surfaced in the Model C during its short lifetime.

Two big shortcomings haunted the Model GP: its lack of power and its three-row design. Neither of these proved very popular with farmers. Nevertheless, the GP was considered an improved version of the Model C and featured the words "General Purpose" on top of the hood.

The early GP's two-cylinder engine was the same as the Model C and displaced 312 cubic inches from a 5.75x6-inch bore and stroke rated at 950 rpm.

The original serial number on the 1927 Model C was 200013, but it was rebuilt at the factory in March 1928 so it could be considered a 1928 model. The new serial number it received on March 28th is 200134. What changes were made to this tractor at the factory remains a mystery.

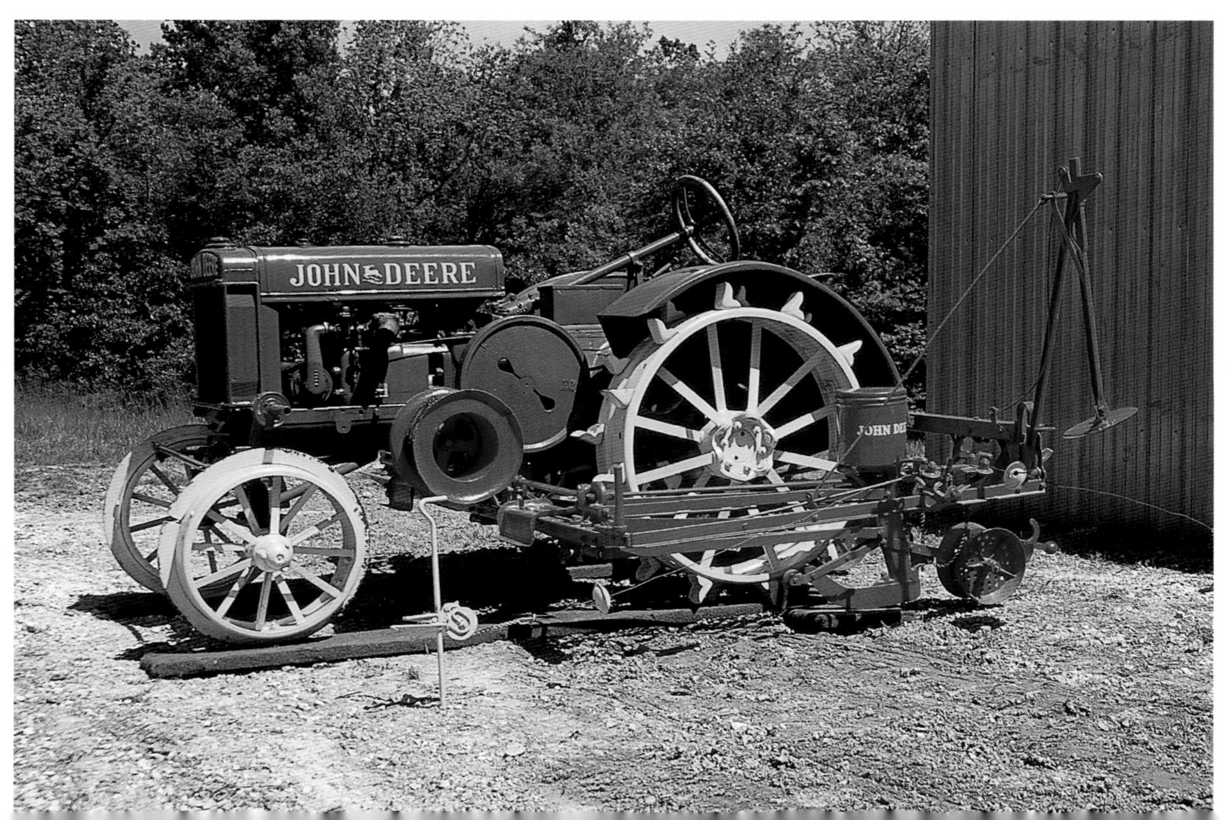

Nebraska test number 153 conducted in 1928 recorded 17 drawbar and 24 belt horsepower operating on kerosene. Although the Model C wasn't tested at Nebraska, the fact that it used the same engine and transmission as the Model GP means these same results likely apply to it.

Butterworth, who had strongly opposed Deere's participation in tractor production, died in 1928, and the presidency of Deere & Company was passed to Charles Deere Wiman, John Deere's grandson. Years later, Wiman would be, in large part, responsible for the decision to end production of the celebrated two-cylinder John Deere tractor.

Deere had a foot in the door with the Model D and was working on a row-crop tractor as the Roaring Twenties drew to a close.

To us, it seems cumbersome, slow, and antiquated, but compared to the horse-drawn check-row planters, this is almost space age. Powered off the Model C's engine, this attachment manages the winding and unwinding of the wire while planting.

This is the Model C outfitted with a Model 301 three-row planter that looks great. According to the owner, the planter worked fine. Farmers, however, never embraced the three-row concept.

In the background is the 1927–1928 Model C with three-row planter. In the foreground is a 1935 Model GP. The GP was one of the first two-cylinders tractors to be equipped with an oilbath air cleaner.

This is a very early 1939 Model H. The option list wasn't a long one for the Model H, and this tractor definitely has the short list—no fenders, wheel weights, lights, or electric starter.

# 3

# 1930s
## The Next Letter

As the Great Depression tightened its grip on the nation and black blizzards boiled across the Plains, Charles Wiman issued a mandate to a small group of Deere & Company engineers. Deere & Company's president insisted that these men design a better row-crop tractor. This was 1931, and Deere's row-crop tractors weren't just faltering—International Harvester Company's (IHC) Farmall was mauling them in the field and on the sales floor.

Meanwhile, Deere had no choice but to continue building the General Purpose tractors already in production while waiting for Wiman's mandate to bear fruit.

### *GP X/O*

The original GP tractors didn't have enough horses under the hood for three-row cultivation, so in May 1930, the crossover GP Standards, sometimes known as the GP X/Os, were introduced. The bore was increased to 6 inches, and the stroke remained at 6 inches, giving a square engine displacing 339 cubic inches. Both the 315- and 339-ci engines were mated to a 3/1 transmission.

Mounted, three-row cultivators and planters were available along with an optional mechanical power lift and 540 rpm power takeoff. Individual rear-wheel brakes were standard equipment. Several options that were eventually offered included rubber tires, an extra-wide front axle, and a lighting package. All Model Cs and early GPs and GPWTs had side-steer design.

The test number 190, conducted in 1931 with distillate fuel, only bumped the results one horse to 18 drawbar and 25 belt. Still, a lack of power coupled with the unpopular three-row configuration made the GP tractors a disappointment to both the company and the farmer. It was no secret to anyone that Deere & Company needed a better row-crop tractor.

As only 68 GP Standard Crossover tractors were produced, they are very collectable.

### *Model GPWT*

Deere finally got it just about right with the Model GPWT, which could handle either two- or four-row planting and cultivating. Beginning in July 1929, the GP Wide-Tread went into production side-by-side with the Standard Model GP. Its 84-inch rear-tread coupled with a tricycle front axle gave Deere its first real contender in the row-crop tractor market.

Like its sibling, the GP, the GPWT began life with the 5.75x6-inch bore and stroke engine. It was followed by the 6-inch bore crossover design, which was replaced with the 6-inch bore with left-side, vertical air-intake stack and right-side exhaust and muffler.

Over-the-hood steering replaced the side-steer design in 1932. Operator comfort was enhanced when the seat location was moved forward over the rear axle

The lettering on top of the hood constantly reminded the owner that he or she was operating a GP Wide Tread. This 1933 model has overhead steering, which means it's one of only 443 (some say 445) copies built.

and raised higher. Other improved features included repositioned throttle and spark controls, a frame that was 6 inches longer, and a tapered fuel tank and hood, which greatly improved operator visibility. To further aid the operator's ability to see while working row-crop, the exhaust and air intake were positioned within the hood line.

Production was slowed during the Great Depression years and ended in October 1933. Only 443 Wide-Treads of the final version with over-the-hood steering were produced.

### Model P

The Model P was a low-production specialty vehicle built for the potato growers in Maine. A Model P was a GPWT with 68-inch rear tread to fit the row spacing of the potato fields. The P-Series model came with 8-inch-wide rear wheels with 16 lugs instead of the 10-inch, 24 lug wheels of the regular GPWT tractors. They carried the same 5.75x6-inch bore and stroke engine as the GP.

The initial production of 150 Ps was built as new units, while the remaining production was rebuilt GP Standard tractors. The P Series was introduced in 1930 and discontinued the same year when inset rear wheels were developed for the GPWT, giving that model the required 68-inch tread for potato farmers.

### Model GPO

Power farming so impressed orchard owners that they also thought it was time to make the switch from horses to horsepower. Orchard growers needed a low-profile tractor to work around and between the trees without damaging the low-hanging branches.

Responding to this need, the Lindeman Power Equipment Company, of Yakima, Washington, modified six Standard Model GPs to produce the initial John Deere Orchard tractors.

Deere liked the concept, and by designing a lower front axle and reversing the rear-axle housing, Deere successfully lowered the vehicle 7 inches. This bit of engineering created the John Deere GPO, which came online in 1931 and stayed until 1935.

Optional fender skirts covered the rear wheels to below the hubs and extended over the flywheel and belt pulley. Radiator guard and curtains were standard equipment. Although the GPO was introduced on steel, rubber tires became an option in 1932.

### *Model GPO-Lindeman*

Following the success of the GPO, the Lindeman Company purchased a number of GPs and fitted them with a crawler undercarriage. These Orchard Crawlers became known as the GPO-Lindeman. How many were built is unclear—probably somewhere between a few and 25 GPOs were converted by Lindeman.

### *Bean Tractor*

The Bean tractor was another special application of the GP Standard model. This time, it was modified to fit the four-row spacing of edible bean and beet growers. Reversing the offset-hub rear wheels and moving them to opposite sides, plus an extended front axle, made a Bean tractor.

In 1931, the parts necessary to convert a Model GP to a Bean tractor became available as an option. Thus, the assembly line Bean tractors were discontinued.

Finally, in 1934, the fruits of Wiman's engineering request became a reality. Deere & Company's long-awaited better row-crop tractor was spelled AANAWAOAOSARANHAWHAIAH—the John Deere Model A.

From this angle, it's obvious that Wide Tread meant, well, "wide tread." The rear tread of 84 inches gave the GPWT the capacity to straddle two rows and accommodate either two- or four-row implements. This was a welcome improvement over the three-row design of the Model C and Model GP.

### *Model A*

Field-testing of the first prototypes of the Model A began in 1931, and by early 1934, engineering was convinced the new tractor was ready.

Production began in April 1934 and closed in 1952 after approximately 300,000 copies had made the journey down Waterloo's tractor assembly line. Variants on the Model A theme included the AN, AW, AO, AOS, AR, ANH, AWH, AI, and AH.

The original Model A was a General Purpose tractor with a two-wheel tricycle front axle. Production of the unstyled vehicle was an impressive 65,978 copies from 1934 to 1938.

Initial production Model A tractors had a 5.5x6.5-inch bore and stroke engine mated to a 4/1 transmission. The two-banger started on gasoline and operated on distillate, was rated at 975 rpm, and displaced 309 cubic inches. According to Nebraska test number

Besides having a brass GP tag, this 1934 Model A is fitted with a beautiful set of French & Hecht dual rear wheels. Did they serve any purpose besides looking great? They must have, but the experts are divided on what that purpose was. Obviously, it would provide more traction and it would add flotation in sandy or boggy conditions.

222, conducted in 1934, the Model A cranked out 18 drawbar and 24 belt horsepower burning distillate fuel. This placed Deere's new tractor solidly in the two-plow class.

Rear-wheel tread could be adjusted from 56 inches to 80 inches and gave farmers the flexibility to work in 40- or 42-inch rows.

An optional hydraulic Power Lift made the Model A one of the most operator-friendly row-crop tractors available at the time. Mechanical muscle rather than operator muscle could now be utilized to raise and lower integral equipment. Still, the engineering department continued to tweak the Model A toward the perfect row-crop vehicle.

In 1937, Deere & Company approached New York's Industrial Designer, Henry Dreyfuss, soliciting his services to add a little style to John Deere tractors. Dreyfuss was intrigued by the prospect of designing an agricultural vehicle and agreed to accept the challenge.

Stylish John Deere tractors began to roll off the assembly line in August 1938 as the 1939 early-styled Model A. Mechanically, the early-styled A was identical to the unstyled A, and, therefore, the horsepower remained the same. Unfortunately, the upgraded Model B could now pull a two-bottom plow about as well as the Model A, and at a significant cost savings. It was obvious to the engineers that the next evolutionary step for the Model A was to kick it into the three-plow class.

### *Model AW*

The unstyled Model AW with a wide front axle didn't appear until 1935, when only eight units were produced for the year. Total production for this variant was only about 303 copies. Of course, the W designates an adjustable, wide front axle underneath the front end of the basic Model A Row-Crop.

Only about 303 of the unstyled Model AWs were manufactured between 1935 through the middle of 1938. Subsequently, it received Henry Dreyfuss styling and became the early styled Model AW. This is a 1936 model.

### *Model AN*

The first year the single front-wheel Model AN was offered was 1935. Only one tractor was made that year, and when production closed in 1938, only 591 unstyled copies of the Model AN were produced.

### *Model ANH*

The unstyled Model ANH was not a true Hi-Crop tractor but rather a Hi-Clearance vehicle that offered an additional 2 or 3 inches of crop clearance under the rear axle. Only 26 copies of the unstyled ANH were ever produced, all during 1938.

"Classic" is a good adjective to describe this Model AW on full steel. Rubber tires were available from the start of Model A production, but when this unit was manufactured in 1936, many farmers were still not convinced that rubber was better.

# Timeline: 1930–1939

**1930**

John Deere Model P: 1930 only.

President Hoover and Congress agree on $116 million emergency job relief.

A drought-relief program targets $45 million for farmers.

December finds 60 branches of the Bank of United States closed. The bank has 400,000 depositors.

**1931**

John Deere Model GPO: 1931–1935.

The president's Emergency Committee for Employment reports that there are between four and five million people jobless.

The Great Depression lingers, causing 800 banks to fail.

**1932**

Riots at Ford Motor Company kill four rioters and injure many more.

Jobless rate in some cities reaches over 50 percent. Two million people are homeless vagrants. Forecast is that 13 million will be jobless by the end of the year.

Franklin D. Roosevelt is elected president of United States.

**1933**

Assassination attempt is made on President-Elect Roosevelt.

President Roosevelt issues emergency executive order, which temporarily closes all the nation's banks to stop the massive "runs" that threaten the entire system.

**1934**

John Deere Model A: 1934–1952.

A severe drought maintains a dusty grip on the nation's breadbasket.

Angry and desperate farmers resist foreclosures.

Countless miles of new roads, thousands of schools, and hundreds of airfields, parks, and playgrounds are built by the CWA.

**1935**

John Deere Model B: 1935–1952.

Approximately 11 million Americans are jobless.

President Roosevelt establishes the Rural Electrification Administration (REA) to help electrify rural America. Only one in ten farms in the nation have electricity.

President Roosevelt signs into law the Social Security Act.

**1936**

John Deere Model Y: 1936 only.

Hoover Dam opens on Colorado River, creating the nation's largest manmade lake.

President Roosevelt wins reelection by biggest margin in history, with 523 electoral votes.

**1937**

John Deere Model 62: 1937 only. John Deere Model L: 1938–1946.

Henry Ford's Ford Motor Company rolls out its 25-millionth automobile.

In his inaugural address, President Roosevelt states, "I see one-third of a nation ill-housed, ill-clad, ill-nourished."

Unions win at General Motors and U. S. Steel, while strikes turn violent and deadly at Ford's Rouge plant and at several steel mills.

Worst floods since 1913 devastate the Midwest.

**1938**

John Deere Model G: 1938–1941.

Unemployment reaches 10 million, and another of President Roosevelt's relief laws is passed—the Emergency Relief Appropriation Act. It provides $3 billion in new funds.

**1939**

John Deere Model H: 1939–1947. Early styled John Deere tractors appear.

Albert Einstein and other leading scientists explain possible military application of atomic power to President Roosevelt.

John Steinbeck's new novel *The Grapes of Wrath* is released.

DDT insecticide becomes the world's first synthetic insecticide.

### *Model AWH*

Like the ANH, the Model AWH was not a true Hi-Crop vehicle but rather a Hi-Clearance model. The extra 2 or 3 inches of clearance under the rear axle resulted from the 40-inch rear tires. The front spindles on the AWH were lengthened to give greater front-axle clearance. There were 27 tractors produced during the only year of production, 1938, for the unstyled Model AWH.

### *Model AO*

In general, the Model AO Orchard and Grove tractors and the standard-tread Model AR tractors received most of the engine, tranny, hydraulic upgrades, and improvements that appeared on the other Model A models.

The unstyled Model AO was the original Orchard version of the unstyled Model AR, which was a derivative of the unstyled Model A. Initially, the Orchard tractor of the A family was an option of the Model AR but soon became a separate model, the Model AO. This vehicle was equipped with low stacks, turning brakes, and side-discharge muffler as standard equipment. Orchard fenders were an option.

Approximately 861 copies were manufactured from 1935 to 36 before it was replaced by the Model AOS.

### *Model AOS*

Specialized is an appropriate word to describe these spinoffs of the Model AR and Model AO. Orchard and grove owners needed a tractor that was short, low, narrow, branch and vine friendly, plus very maneuverable. That's just what Deere gave them with the Model AOS Streamlined, which was introduced in 1937.

Orchard and grove owners responded enthusiastically by purchasing 512 copies during the first year. Then, for reasons not totally understood, they just as quickly abandoned the AOS and sales fell every year until production was cancelled in 1941. The tally totaled 820 copies before it gave way to the reincarnate Model AO.

### *Model AR*

Deere's answer to the call for a standard-tread two-plow tractor for open-field drawbar work was the unstyled Model AR. It was a vast improvement over the Model GP, and 17,384 copies of the AR were made and sold from 1935 to 1949. The Model AR had the distinction of being the last unstyled John Deere tractor in production.

### *Model AI*

A derivative of the Model AR, the unstyled Model AI had many features that were only found underneath yellow paint and black, stenciled letters. One such feature is a front axle with bosses and notches. Compared to the Model AR, this special axle is set back 7 inches, giving the AI a shorter wheelbase and a shorter turning radius. Also, the drawbar was strengthened and designed to allow for vertical adjustment.

Some of the options and special equipment included solid rubber tires, special seat mount, front- and rear-wheel weights, low-pressure rear tires, side power shaft, battery and electric lighting, plus an overdrive assembly. These Industrial vehicles could be fitted with a street flusher and sprinkler, a crane, or a snowplow. Manufactured from 1936 to 1941, the Model AI didn't have its own serial numbers. Instead, they are intermingled with the AR and AO numbers.

It's difficult to imagine a more inopportune time to introduce a new line of tractors, considering the conditions in the nation's Great Plains. Despite a depressed economy and black blizzards, Deere forged ahead with its new tractors. It proved the right thing to do, as the Model A and the Model B were destined to become phenomenal successes.

### *Model B*

The Model B, Deere's smaller version of the extremely popular Model A, featured adjustable wheel tread, individually controlled rear-wheel brakes, and four forward speeds. Variations included the BN (single

A nicely restored 1937 Model D includes a couple of the owner's custom features. The modifications are explained in subsequent captions.

front wheel), the BR (standard front axle), the BO Lindeman (Orchard Crawler), the BW-40 (Narrow Row-Crop), and the BW (adjustable, wide front axle), and BWH-40 (narrow row-crop). Over 306,000 Model Bs sold during its manufacture from 1935 to 1952. These stellar numbers make it the highest-volume production tractor in Deere's history. The Model B's widespread use resulted in many versions before the Model 50 replaced it in 1952.

Initially, the Model B's valve-in-head, two-cylinder engine was a 4.25x5.25-inch bore and stroke rated at 1,150 rpm. Because the small engine had only 149 cubic inches, no decompression cocks were necessary. Put through the paces at the Nebraska test lab in 1935 as test number 232, the Model B read 11 drawbar and 16 belt horsepower. A separate gasoline tank allowed the Model B to start on gasoline and switched to lower-cost kerosene, tractor fuel, or distillate once the engine was warmed up. A four-speed transmission was standard from the B's introduction until 1941.

The first unstyled Model Bs are now referred to as four-bolt tractors due to the four bolts securing the front pedestal. Another design characteristic of the very early Bs is the centerline fuel-tank filler caps that are found on approximately the first 509 Model Bs.

Model B tractors built from their introduction in 1935 until 1937 are referred to as short frames. During 1937, the frame was lengthened 5 inches to accommodate mid-mounted equipment designed for the Model A tractor. The long frame resulted in several changes to the Model B's look and function. Affected were the hood, fuel tank, steering shaft, exhaust pipe, carburetor intake, and drawbar.

John Deere's hydraulic power lift was first introduced on the Model A in early 1934 and became available on the Model B when production started late in 1934 (considered 1935 models). This was only a lift-and-lower hydraulic system, as the highly advanced Powr-Trol wasn't available until August 1945. The Powr-Trol was much more than a lift-and-lower system, for it provided selective control of the rockshaft, or a double-acting remote cylinder.

Electric starter and electric lights also became available as an option in 1939. The 7-inch headlights were the same option as offered on the Model D. Rubber tires, too, were offered as an option, from the very start of production. Model B rear-tread width was adjustable from 56 to 80 inches.

Industrial designer Henry Dreyfuss completed styling for the Model B in 1937, and the new look was introduced in fall 1938. Along with the new styling, the bore and stroke was increased to 4.5x5.5, which upped the displacement to 174 cubic inches.

Nebraska test number 305, conducted in 1938 on a Model B with the new, larger engine, pegged the results on steel at 14 drawbar and 18 belt horsepower. Running on rubber, the numbers were 16 drawbar and 18 belt horsepower. The tests were conducted using distillates as fuel.

### *Model BN*

Having only a single front wheel, the Model BN was first advertised as a garden tractor designed for vegetable growers who planted in rows 28 inches or less. The Model BN shared the same specs as the Model B and underwent the same engine and transmission changes during its lifetime. Approximately 1,001 unstyled Model BNs were built, 24 of which are the now-rare four-bolt tractors.

### *Model BW*

Another first-year production Model B variation was the unstyled Model BW fitted with an adjustable, wide front axle. By the time the Model BW was introduced, the four-bolt pedestal had been replaced with the eight-bolt design. Engine and powertrain specs were the same as the basic Model B.

In 1936, six Model BWs were modified to produce a standard-tread tractor that could work row-crop planted in 40-inch rows. Both the front and rear axles on these Model BW-40 vehicles were shortened to produce its very narrow tread for specialized row-crop. As these vehicles were created from the Model BW, the engine and drivetrain shared the BW stats.

### *Model BNH*

Late in 1937, greater clearance was provided for the Model BN by using larger wheels and tires. The standard front wheel was replaced with a 16-inch rim, and the 36-inch rear rims where changed to 40-inch. The larger wheels and rubber tires elevated the tractor's stance enough to be called a Hi-Clearance machine. The BNH was discontinued in 1946 and, therefore, it wasn't found with the late styling.

### *Model BWH*

The same changes that applied to the Model BNH changed the Model BW into a Hi-Clearance, Wide-Tread Row-Crop tractor. Greater rear-axle crop clearance was gained by using the same rear axle as the BNH. A modified BW front axle with longer

Three features of the Model 62 are apparent when viewed from the backside: the "JD" cast in the differential case, the serial number plate, and the pipe frame. This particular tractor is serial number 62-1048.

spindles was used to increase clearance under the BWH's front axle.

The BWH received the early styling but not the late styling as it, too, was discontinued in 1946.

### *Model BWH-40*

Following the same pattern as the BW-40, the Model BWH-40 was fitted with a narrow rear and front axle. Add the larger rear tires and the longer front spindles, and it's a Hi-Clearance machine for bedded crops in 20- or 40-inch rows. Clearance under the front axle was 22.875 inches. An unknown number of these now-rare models were built, and they weren't offered as a late-styled tractor.

The various versions of the Model B provided Deere with good representation in the row-crop market, but it still lacked a small standard-tread tractor to help round out the B Series line.

### *Model BR*

This standard-tread vehicle, the Model BR, proved to be an excellent primary tractor on small farms and a useful secondary tractor for large operations. Production began in the summer of 1935.

The Model BR started life with the same engine and transmission as the other B tractors, but the BR's muffler was an orchard-type, which allowed the exhaust to be directed upward, downward, or horizontally.

In June 1938, the 4.25x5.25-inch bore and stroke engine was replaced with the larger 4.5x5.5-inch bore and stroke design, providing 174 cubic inches. Cylinder decompression cocks were now necessary and became standard on the larger engine.

Because the BR's 68-inch wheelbase was shorter than other B models, it required the fuel tank to be shortened and widened to allow approximately 12 gallons of fuel capacity. The BR was available with a wide selection of tire and wheel options in either steel or rubber. The customer could mix-and-match steel and rubber.

It required only a few changes to turn a Model BR into an Orchard model. Shields over the air intake, gasoline tank cap, and fuel tank cap—plus optional full citrus fenders that covered the rear wheels to below the axle—prevented damage to tree limbs.

The BR and BO were never styled by Henry Dreyfuss and his team of industrial designers.

### *Model BO*

The Orchard version of the BR was a respectable seller with 5,083 copies produced from 1935 to 1947. Shields over the air intake, gasoline tank cap, and fuel tank cap prevented limbs from catching on the tractor when working close to trees or vines. In June 1938, the 4.25x5.25-inch bore and stroke was increased to 4.5x5.5 inches. This vehicle was never styled by Henry Dreyfuss.

### *Model BI*

One of the changes that transformed a Model B into an Industrial Model BI was yellow paint and black lettering. These yellow John Deere versions were manufactured from 1936 to 1941.

Although the Model BR and BI had major components in common, there were other changes, such as a front axle set back 5.25 inches, a beefier drawbar, larger outer bearing on the rear axle, a padded seat, and a lower air stack. If the BI was equipped with rubber tires, individual rear breaks were standard.

The Model BI didn't have separate serial numbers. Its serial numbers fall within the standard-tread Model B numbers.

The 1937 Model 62 is a product of the John Deere Wagon Works in Moline, Illinois. There were only 78 copies ever produced. The belt pulley is the only option on this tractor.

### *Small Letter Tractors*

The first John Deere tractor targeted at small-acreage farms, such as truck gardens and vegetable growers, was the Model Y. The prototype was commissioned in 1936 and built at the John Deere Wagon Works in Moline, Illinois.

Deere & Company's bottom line for 1931 was a $1 million on $27.7 million in sales. The following year, the loss was $5.7 million on $8.7 million in sales. The year 1933 saw sales rise a bit to $9 million but with a $4.3 million loss. However, management insisted that R&D continue despite the rough times. This foresight eventually resulted in the highly successful Model A and Model B tractors, which were introduced in 1935.

Their success was not yet established in 1935 when the engineering department was once again asked to develop another tractor to fill a specific niche in the line. So, considering the company's financial woes of the past years, its no wonder that the budget was tight at the Wagon Works new tractor program.

### *Model Y*

In 1936, with the modest budget in mind, the Model Y was built from existing Deere parts and outside vendor components. A Novo vertical two-cylinder gas engine was the first powerplant, and it was soon replaced with a two-cylinder Hercules engine. A Ford Model A transmission and steering gear were used along with wheels from a John Deere manure spreader. Probably all 24 (or 26, depending on the source) of the Model Ys—with the possible exception of one tractor—were recalled by the factory and destroyed.

The 3x4-inch bore and stroke engine displaced 56.48 cubic inches, and although it was never tested at Nebraska, it probably hit the scale at about 8 drawbar horsepower.

A Garden Cultivator and a Garden Planter were implements designed specifically for the Model Y. All Model Ys were built during 1936.

### *Model 62*

In early 1937, the Y was redesigned and renamed the Model 62, which served more as a prototype for the Model L than as a consumer product. There were either 78 or 79 copies of the Model 62 produced, all in 1937.

The Model 62 was an unstyled tractor with a large "JD" in the casting below the radiator. The "JD" is also found on the back of the differential casting. These vehicles were powered by a 3x4-inch bore and stroke Hercules NXA vertical gasoline engine displacing 56.48 cubic inches. The cooling system was typical John Deere thermosiphon.

It was never tested at Nebraska but probably had the same 7 or 8 drawbar horsepower as the unstyled Model L.

### *Model L*

When the Model L began to roll off the assembly line in late 1937, it was Deere's smallest farm tractor. It was obviously a descendant of the Model Y and 62, but there were some noticeable changes. The attractive "JD" on the front shield below the radiator was gone, the fenders were changed to a clamshell design, and the cast-iron wheels were changed to steel.

Approximately 1,502 Model Ls in unstyled configuration were manufactured. A small number of the unstyled Model Ls were painted industrial yellow and served outside of any agricultural application. In 1938, the Model L was Dreyfussized to match the Model A, Model B, and Model D tractors. The styling was achieved primarily by new tinwork on the hood and grille. The styled Model L continued in the Deere General Purpose line until 1946 and racked up total production numbers of 11,225 copies.

A taller front axle and larger rear tires turned the Model L into a Hi-Clearance model, a transformation that added approximately 3 inches of clearance. All 3,263 of these vehicles used the Hercules engines.

A Deere-designed, Hercules-built NXA vertical L-head two-cylinder gasoline engine was the powerplant.

The space between the top radiator tank top and the steering shaft helps identify this as a low-radiator Model G. Approximately 3,200 copies were produced before the radiator and cooling system were improved to solve an overheating problem. Deere & Company initiated a program to correct this problem on tractors already sold, but a few, like this one, escaped.

With 3.25x4-inch bore and stroke, it displaced 66 cubic inches and turned at a rated rpm of 1,550. The 3/1 transmission had forward speeds of 2.50, 3.75, and 6 miles per hour. Nebraska's 1938 test number 313 verified that the Model L produced 9 drawbar and 10 belt horsepower.

The unstyled and early-styled Model L era ended in 1941.

### Model G

The next model in the Deere line of lettered tractors was to be the Model F. However, International Harvester fielded its Farmall F first. To avoid any confusion or conflict, Deere's new offering became the Model G. The Model G filled the role of big row-crop tractor in Deere's stable from 1937 until the Model 70 replaced it in 1953. Production of the unstyled Model G began in May 1937 and ran until the end of 1941, resulting in 10,677 copies.

Power for the Model G came from a 412-ci, two-cylinder engine with 6.125x7-inch bore and stroke. This powerplant and a 4/1 transmission put the G in the three-plow class. However, the cooling system quickly proved inadequate for the large engine, and the radiator capacity was increased to alleviate the problem. The change required that the hood and fuel tank be modified to accommodate the larger radiator. Today, Model G tractors with the small radiator are referred to as low-radiator Gs by collectors. The unstyled Model G was only offered with the two-wheel tricycle front axle.

Nebraska test number 295 conducted in 1937 on an unstyled Model G operating on distillate fuel netted 27 drawbar and 35 belt horsepower.

### Model H

Max Sklovsky, Deere's chief engineer in Moline, and staff had been working on small-tractor concepts since early in the 1930s. This engineering department designed the Model Y, the Model 62, and the Model L between 1935 and 1937.

The smallest member of the General Purpose John Deere line was the Model H, introduced in 1939. It's the smallest if you don't count the Model L or LA. It was never an unstyled tractor, and four versions were available during its lifetime: The H, HN, HWH, and HNH.

The build date on this Model G was March 30, 1938. It was then shipped to the Kansas City branch on March 31, 1938. Its serial number is correct for a large-radiator Model G. The restorer said that the tractor was the cleanest he'd ever seen and needed very little mechanical overhaul.

Sometime in 1937, the Waterloo facility began work on its own small experimental tractor that resulted in the Model H. Deere & Company Decision No. 7,900, dated September 29, 1938, defined the role envisioned for the Model H tractor.

> To meet the needs of small farms for limited power at low cost and for supplemental power on large farms, we will authorize the production of a lighter General Purpose Tractor designated as the Model H with the following features:
>
> I. Low production cost, obtained by:
>
> A. Low material cost resulting from extensive use of high-strength materials and simplification of design with a net dry weight of approximately 2,070 pounds.
>
> B. Low labor cost secured through reduction in total number of parts and in simplification of parts.
>
> II. Low operating cost, which results from:
>
> A. Low fuel consumption due to power requirement closely approximating the economical load range of the engine.
>
> B. The use of a two-cylinder variable speed engine adapted to low-cost fuels.
>
> C. Decreased rolling resistance due to light weight, large diameter wheels and the use of rubber tires on rear and front wheels.

The Model H fit between the Model L and the Model B, offering customers a little less power and a little less cost than the Model B with a little more power and a little more cost than the Model L. The prototypes, or experimental tractors, carried the designation XO and were unstyled, bare-bones vehicles. Extensive field-testing during the summer of 1938 verified that the H tractor was ready for production. The production line began to roll in December 1938.

A Deere-built, two-cylinder horizontal engine with 3.562x5-inch bore and stroke displacing 100 cubic inches powered the Model H. It was an all-fuel engine cooled by thermosiphon, and it was wound a little tighter than other Deere engines, with a rated speed of 1,400 rpm.

This Waterloo two-cylinder horizontal engine design was a bit different as explained in Decision No. 7,900.

> The transmission is of the all-spur gear-type, employing but three steps of reduction. The design differs from that of the Models A, B, and G in that power is taken from the camshaft of the engine rather than the crankshaft, thereby taking advantage of the 2 to 1 speed reduction from the crankshaft to camshaft, and affords quiet operation due to the low velocity of the belt pulley gear. The use of this three speed transmission with the variable speed engine provides a wide range of speeds adapted to all farm purposes.

A Model H definitely wasn't a muscle tractor, as it could muster only 12 drawbar and 14 belt horsepower during Nebraska test number 312, performed in 1938. During this test, the Model H set a fuel-economy performance record of 11.95 horsepower hours per gallon of distillate.

Fully outfitted with rubber tires was the only way a customer could buy any Model H tractor. When introduced, the Model H standard rear tires were 6.5x32, but 7.5x32 tires were offered as an option. By March, the 7.5x32 had been adopted as the standard equipment rear tire.

This tractor's low price tag and low-cost operation placed it within the budget of approximately four million small farmers who farmed 80 acres or less, and 57,450 purchased a Model H. An additional 978 purchased an HN; 126 bought the HWH; and 37 took home a HNH. No separate serial numbers were assigned to the HN, HNH, and HWH.

At first glance, this appears to simply be a nice, but ordinary, Model A. It is a model A, but ordinary it is not. This 1942 tractor has a special one-of-a-kind feature.

# 4

# 1940s

## Deere is Letter Perfect

With the arrival of the 1940s, Americans were able to look back knowing that they had survived the Great Depression and the Dirty Thirties, and they thought that the present and future surely must hold better days. The nation moved on with the business of life.

## Moving Forward

In 1940, the business of tractors at Deere & Company was built around the Model D, Model A, Model B, Model G, Model H, and Model L. It wasn't a lineup made in heaven, but it held promise of better things to come. As the new decade rolled along, improvements were continually incorporated into the models.

## Model L

In 1941, the Model L was given a 3.25x4-inch Deere-designed, Deere-built vertical gasoline engine. The Model L never offered a PTO, but instead, a belt pulley was available as an option. Electric lights and an electric starter were an option later in production.

### *Model LI*

Official production of the LI, the Industrial version of the Model L, began in 1941. However, some earlier production L tractors were considered Industrials.

The Deere-built engine was standard in the Model LI from 1941 to 1946. Any Model L styled tractors considered Industrials before 1941 were fitted with the Hercules-built engine.

### *Model LA*

Under the hood of the Model LA was the Deere-designed, Deere-built vertical two-cylinder engine with 3.5x4-inch bore and stroke that displaced 77 cubic inches and rated at 1,850 rpm. That's a 0.25-inch greater bore, a 300 rpm increase, and 11 more cubic inches than the Model L, which all added up to a 40 percent increase in power. Now, the LA was churning out about as much horsepower as the Model H. An unknown number of the industrial verion of the Model LA was produced, although approximately 12,475 copies of the Model LA were turned out at the Moline Wagon Works—not bad for a bunch of carpenters and wheelwrights.

Options included an electric starter, lights, generator and battery, adjustable front axle, PTO, front and rear weights, and belt pulley.

The Model LA showed 13 drawbar and 14 belt horsepower in test number 373 conducted at the Nebraska testing facilities in 1941.

Production of all variations of the Model L and LA ended in 1946. At this time, Deere & Company's tractor line included the Model D and Model G as the big tractors; the Model A, Model B, and Model H as mid-line power; and the small end was covered by the Model L and Model LA.

At last, there wasn't a farm anywhere that couldn't find just the right size of John Deere tractor. However, big changes were on the way.

The owner of this 1945 Model LA bought it from a friend whose father was the original owner and lived in Table Rock Lake, Missouri. The engine was in bad shape, and the restoration "took a lot of money," according to the restorer. The result, though, is a keeper.

## Model A

One such change for the Model A was initiated in 1940. The stroke was increased from 6.5 to 6.75 inches, while the bore remained at 5.5 inches. This upped the displacement from 309 to 321 cubic inches. These 12 extra cubic inches coupled with an improved cylinder head and pistons bumped the drawbar horsepower from 18 to 26, as testified by the 1939 Nebraska test number 335. Belt horsepower rose from 24 to 29.

When the early-styled Model A was introduced in 1938, it retained the four-speed transmission—a beefed-up version to handle the extra power. However, by now farmers wanted even more gears and higher speeds, so with the 1941 Model A vehicles equipped with rubber tires, the customer could have a six-speed tranny. The tractors on steel retained the four-speed tranny.

Changes in operator convenience included optional items like a factory-installed electric starter and lighting package, and the new Powr-Trol hydraulic system that became available in 1945. This system could operate the rockshaft as well as a remote cylinder.

World War II greatly impacted the Model A (as it did most tractor models from most manufacturers), and production was suspended for several months in late 1942 and early 1943.

Despite the war, the early-styled Model A, including all variants, racked up about 103,294 copies before it bowed out to the late-styled model.

In 1947, electric lights and an electric starter became standard equipment on what are referred to as the late-styled Model A. The starter was enclosed to protect it from the elements, and by moving the starter location to the bottom of the main case, the flywheel could also be enclosed.

Dreyfuss proved that comfort and function could exist simultaneously when he moved the battery location to the seat-support box under the new foam-padded seat. The move accommodated the two 6-volt batteries required for the 12-volt system that debuted with the late-styled package.

The new foam-cushioned seat with a cushioned backrest advanced operator comfort light years beyond the old metal pan seats. The new seat coupled with the optional Roll-O-Matic front-pedestal suspension for Tricycles added yet another welcome advancement in operator comfort and convenience.

If the buyer wanted the new Powr-Trol, he could specify it as an option. However, the hydraulic Power Lift was standard equipment on all Model A tractors.

Perhaps the biggest change on the late-styled Model A was found under the hood. This change demanded a big adjustment concerning Deere & Company's long-standing commitment to low-grade fuels.

For years, Deere's two-cylinder engines were designed specifically for burning kerosene, tractor fuel, and distillates. Deere heavily promoted the economic advantages of this low-grade, low-cost fuel, even advertised that the fuel savings alone could pay for the tractor.

This all began to change as petroleum technology advanced during and after World War II. Now, the refineries were wringing more gasoline out of a barrel of crude oil and less of the heavy fuels like kerosene and distillates. Consequently, the price advantage of heavy fuels began to erode quickly.

Gasoline also tolerated higher compression and yielded more horsepower per unit of fuel. For these reasons, Deere made a decision to offer a gasoline engine for the late-styled Model A vehicles.

When the Model A fitted with the new gasoline engine was tested in Nebraska test number 384 in 1947, it posted 34 drawbar and 38 belt horsepower.

### Model AW

The early-styled Model AW production reached 1,849 copies before it bowed out to the late-styled version in 1947.

Production numbers were much better for the late-styled Model AW, with 6,275 vehicles manufactured between 1947 and 1952.

### Model AN

In 1947, all of the upgrades and improvements incorporated in the Model A found their way to the final version of the three-wheeler—the late-styled Model AN. Once again, gasoline proved more popular than all fuel for powering the Model AN—4,011 gasoline tractors and 1,212 all-fuel vehicles. Production closed in 1952.

The Model A, Model AN, and Model AW could be purchased outfitted with a bewildering array of rubber tires or steel wheels—or a combination of rubber and steel.

### Model ANH

Material restrictions caused by World War II halted production of the early styled Model ANH in 1943, and there wasn't a single year in which more than 46 of these tractors were built. Consequently, the total production only accounted for a total of 236 copies

Regardless of the shortcomings this 1948 Model M and loader might have, it beats using a pitchfork when you need to move a mountain of manure.

made between 1939 and 1947. Because of low demand, the ANH wasn't produced in a late-styled version.

### *Model AWH*

Production figures show only 361 copies of the early-styled AWH tractor were built, all between 1939 and 1947. The Model AWH, like the Model ANH, didn't make the transition to the late-styled version.

### *Model AO*

Production of the model AO resumed in 1941 and ran through 1949. During that time, 1,825 copies of the Model AO were manufactured.

### *Model AR*

In 1941, a Model AR, operating on distillate fuel, was tested at Nebraska, test number 378, and recorded 26 drawbar and 30 belt horsepower. Tested again as test number 429 in 1949, this time as a gas burner, the results were 34 drawbar and 37 belt horsepower.

Equipped with the new, improved two-cylinder gasoline engine, the now-stylish Model AR was comfortable with a three-bottom plow in most any soil and could handle a four-bottom in light or sandy conditions. The all-fuel version was a two-three plow outfit.

## Model B

In 1941, two additional forward gears were added to the rubber-tired Row-Crop Model B. The six-speed tranny was a three-speed with a second gear-shift lever providing a high-low side to the three basic gears.

However, during the war years, when rubber was in short supply, many Model B tractors were produced on steel wheels. In this case, the two extra top-end gears were not added to the transmission. Because farmers were getting used to doing farm work faster, many owners added these two extra gears when their tractors could finally be fitted with rubber tires.

Tests on a Model B with the new 6/1 transmission began at Nebraska in late 1940, but weren't finished until the spring of 1941. So, test number 366 is listed as a 1941 test. With distillates as fuel, the results were 18 drawbar and 20 belt horsepower.

### *Model BO-Lindeman*

Take one John Deere Model BO chassis and marry it to a Lindeman track assembly, and you end up with a cute little crawler.

From 1939 to 1947, production of Model BO Crawlers from the Lindeman plant totaled 1,645 units plus 29 built from Model BRs and one converted from a Model BI.

## Model G

Styling and war restrictions came to the Model G at about the same time. While styling gave it the family look of the Model A, B, and D, war restrictions denied the company a price increase to offset the cost of this new styling.

### *Model GM*

In 1942, to circumvent war restrictions, Deere gave the tractor a new model designation—GM. Then, after only eight months, Model GM production was stopped and didn't resume until October 1944.

When rubber tires weren't available, the Model GM was shipped on steel, which required that fifth and sixth gears be deleted from the tranny. Many, if not most, of these tractors were retrofitted with the top two gears once rubber tires became available after the war.

Options offered were electric starting, electric lighting, or both; Power Lift; fenders; and Powr-Trol.

Like the unstyled Model G, the Model GM was a Tricycle only—no wide front, no single-wheel front, no Standard front tread. When production ended in March 1947, approximately 8,764 GMs had been manufactured.

### *Postwar Model G*

In March 1947, the Model G returned as an early-styled vehicle that was identical to the Model GM except for the serial number plates that said "Model

G" rather than "GM." Production tallied 2,403 copies. Early-styled Model Gs had a production run only slightly longer than the life expectancy of a Great Plains grasshopper. The new, improved late-styled Model G was launched in July 1947 for the 1948 model year. Production numbers for the late-styled Model G weren't awesome, but were respectable with 31,913 tractors produced.

Electric starting and electric lights were standard equipment on the new G. Even better, the old iron seat was replaced with a new padded seat with back rest. Even better yet, the Powr-Trol was standard equipment.

There was significant change to the cooling system in 1952, which came about due to the copper shortage precipitated by the Korean War. Steel radiator cores don't dissipate the heat as well as copper, so use of a water pump was adopted.

In 1947, ten years after the Model G was first tested, test number 383 rated 34 drawbar and 38 belt horsepower using tractor fuel.

### *Model GN*

Converting the Model G to a GN was easily accomplished thanks to the two-piece, or split, pedestal that Deere had perfected. Once the single wheel was under the front, it was a matter of deciding if the rear tread was adequate. If not, a rear axle was available that would provide tread width up to 104 inches.

Except for the single front wheel, the GN was a basic Model G under the hood and throughout the powertrain. It shared all of the standard and optional equipment of the Model G. The early-styled Model GN was the customer's choice only 49 times.

A total of 1,522 copies of the late-styled Model GN rolled down the assembly line at Deere's tractor plant in Waterloo, Iowa, from 1947 to 1953.

### *Model GW*

The two-piece, or split, front pedestal made it easy to accommodate any front wheel combination a customer wanted.

This 1942 Model GM was originally shipped to the Kansas City, Kansas, branch and then to Salina, Kansas. In the following 60 years, it hasn't been more than 100 miles from the Salina dealership that first sold it.

The late-styled Model GW was more in demand than the late-styled Model GN, with 4,666 copies needed to meet customer demand. The wide adjustable axle provided a tread width from 56 to 80 inches.

## Model H

Hydraulic power became available for the Model H tractors in late 1940, after which the hydraulic power lift could be a factory or field installation.

Some optional equipment for the Model H included an electrical system for starter and lights, or just a

# Timeline: 1940–1949

**1940**

U.S. population surpasses 131 million.

One-third of all farms now have electricity.

President Roosevelt wins an unprecedented third term.

Germany begins air strikes against England.

President Roosevelt prepares American industry for a military buildup.

**1941**

John Deere Model LA: 1941–1946.

General Motors manufactures 50 percent of all automobiles produced in the United States.

Japan attacks Pearl Harbor. America declares war.

**1942**

John Deere Model GM: 1942–1947.

Sale of new autos and trucks is banned by the Office of Production Management.

Sugar and gas rationing goes into effect. U. S. forces surrender on Bataan. It is a prelude to the Bataan Death March.

First sustained nuclear chain reaction is achieved by the United States.

Deere & Company President Charles Wiman serves in the U.S. Army from 1942 to 1944, during which time, Burton F. Peek is acting president.

**1943**

All wages and prices are frozen by President Roosevelt.

U.S. troops are finally victorious on Guadalcanal.

War production at home hits high gear, with the B-24 factory at Willow Run, Michigan, producing 86,000 planes in 1943.

**1944**

The scrap effort has contributed one-half of the steel, tin, and paper required by the war effort.

Allied troops storm ashore on D-Day. Meat rationing ends.

President Roosevelt wins a fourth presidential term. Harry S. Truman is the new vice-president.

**1945**

Manhattan Project tests atomic bomb. The top-secret project employed 105,000 men and women, most of whom didn't know what they were helping create.

Henry Ford steps down as head of Ford Motor Company.

President Franklin D. Roosevelt dies. Truman becomes the 32nd U.S. president.

German leaders sign an unconditional surrender on May 7. Atomic bomb is dropped on Hiroshima, Japan, on August 6 and Nagasaki, Japan, on August 8. Japan surrenders, ending World War II.

**1946**

John Deere Model M: 1947–1952.
The Soviet's Iron Curtain clamps down across Europe.

U.S. workers (4.5 million of them) strike.

**1947**

John Deere Model G: 1947–1953.
Henry Ford dies.

**1948**

President Truman defeats Thomas E. Dewey in election.

Richard and Maurice McDonald open hamburger joint. Hamburgers sell for 15 cents, French fries for 10 cents, and milkshakes for 25 cents.

**1949**

John Deere Model R: 1949–1954.

General Motors announces profits of $500 million in first three quarters.

starter or just lights. A radiator shutter was offered instead of the standard curtain, as was a PTO shaft with shield. Fenders were also an option.

Production of the Model H Tricycle accounted for 57,363 copies.

### *Model HN*

The sales department recognized the need for the HN shortly after production began on the H tractor. Officially, the Model HN was launched by Decision No. 9,000 dated January 15, 1940. The first HN was produced on February 19, and approximately 1,077 units were built.

### *Model HWH*

The first Model HWH rolled off the production line on March 5, 1941. Thereafter, only 126 copies of the Model HWH tractor were manufactured on the Waterloo assembly line.

Two different front axles were available for the Model HWH. The narrow center section allowed the front wheels to be set for crops planted in narrow rows. The wide center section front axle allowed wider tread settings for crops planted in 42-inch rows.

These tractors will regularly be equipped with a front axle (short) to give tread range from 40 to 52 inches by increments of 4 inches. A special equipment front axle (long) and drag links will be adopted to give tread range of 56 to 68 inches by 4 inch increments. This can be assembled on the tractor at the factory or furnished separately to users who may desire the full tread range from 40 to 68 inches. The rear-wheel tread with the 8-38 tires is same as for Model H—44 to 84 inches.

### *Model HNH*

Only 37 copies of the HNH were produced, making it by far the rarest of the H models. The first Model HNH tractor was produced on March 11, 1941.

World War II interrupted production of the Model H series several times, sometimes for as long as 12 months. No HNH or HWH models were produced after the close of World War II.

Deere & Company invested over $1 million to develop the Model H, and it seemed like fate had conspired against Deere's small-tractor program. Actually, it wasn't fate. Rather, it was Henry Ford and Harry Ferguson who plagued the Model H, the Model L, and the Model LA.

Deere introduced the H in 1939, which happened to be the same year that Ford and Ferguson showcased their revolutionary new Ford 9N with the Ferguson hydraulic system and three-point hitch. Sales numbers on these little gray tractors were what the rest of the industry could only dream about. During the war years, when many manufacturers, including Deere, discontinued production of many models, the new Ford-Ferguson 9N and 2N racked up sales of almost 200,000 units by the end of the war in 1945.

A close up of this 1942 Model GM shows the offset of the frame, which is a signature feature of the Model G and Model GM.

It was soon obvious that Deere would have to meet, or exceed, the Ford-Ferguson features, such as electric lights, electric start, low cost, and advanced hydraulics to handle integral, or mounted, equipment. This had to be done quickly if Deere was to be competitive in the small-tractor market.

Fortunately, just such a tractor was on the drawing board at Deere & Company.

Electric start, lights, PTO, and fenders are the goodies on this 1942 Model GM. Its serial number is 13075, with a build date of March 9, 1942, and ship date of March 11, 1942. The destination was the Kansas City branch. It left the factory on full steel—skeleton rear and 24x5 front.

## Model M

A replacement for both the L and LA, and hopefully some competition for the Ford-Ferguson, the Model M General Purpose Utility tractor was introduced in 1946 and was Deere's smallest tractor.

The Model M was a two-plant tractor. Designs and prototypes were developed at the Moline Tractor Works facility in Moline, Illinois, but actual production occurred at the new Deere & Company plant in Dubuque, Iowa.

The Model M differed in several aspects from Waterloo-built John Deere tractors. The engine was two-cylinder, but it was vertical, not horizontal, and these powerplants didn't have the hallmark Johnny Popper external flywheel or a hand clutch.

The newly designed two-cylinder vertical gas engine had 4x4-inch bore and stroke, displaced 101 cubic inches, and operated at 1,650 rpm.

Test number 387 conducted at Nebraska in 1947 verified that the Model M developed a modest 18 drawbar and 20 belt horsepower.

Although the Quik-Tatch hitch and implements were a worthy competitor to the Ford-Ferguson three-point system, the M never enjoyed the sale numbers of the Ford. When all versions of the Model M are tallied, total production was approximately 87,816 copies when production ended in 1952.

### *Model MT*

Introduced in 1949, the MT was a row-crop version of the Model M Utility tractor. Three front-axle versions were available, including the dual-wheel Tricycle, the wide adjustable, and the single front wheel.

The engine and drivetrain for the Model MT were the same as the Model M, and when tested at Nebraska in 1949, test number 423, the results were 18 drawbar and 20 belt horsepower—identical to the Model M. The MT sold in very respectable numbers, with 30,473 copies completed before it was discontinued in 1952.

### *Model MI*

By redesigning a new lower front axle and rotating the final drive housings, Deere engineers lowered the Model M some 8 inches and reduced the wheelbase an equal amount. These were some of the obvious changes that produced the MI, or Industrial, version of the Model M.

The total number of M vehicles wearing highway orange or industrial yellow with black decals was 1,032 from 1949 to 1952.

### *Model MC*

The Model M Crawler was also introduced in 1949. Produced in Dubuque, Iowa, it differed from the previous John Deere crawlers in that it didn't use the Lindeman undercarriage.

The tracks and undercarriage for this vehicle were designed by Deere engineer Harold Borsheim at the Dubuque facility and built at the Dubuque plant.

Tested at Nebraska in 1950, the Crawler posted 17 drawbar and 21 belt horsepower in test number 448.

The Model M Crawler was duplicated 10,510 times on the Dubuque assembly line before production stopped in 1952.

## Model R

Rudolph D. Diesel perfected the diesel engine in 1892, and by 1900, diesel engines had found their way to the United States. However, it wasn't until 1931 that Caterpillar began using this new power source in its crawlers. Two years later, International Harvester designed the first commercially successful diesel-powered wheel tractor.

Deere & Company's experimentation with diesel goes back as far as the 1930s, but it wasn't until the 1940s that a serious effort was mandated.

The Model R was conceived during Deere's protracted dilemma concerning the future of the two-cylinder engine. Besides the two-versus-more cylinder debate,

The current owner of this 1940 Model B Slant Dash was a 17-year-old hired hand when the farmer brought it home to replace the workhorses. The hired hand finally located the tractor and was able to purchase it about five years ago. Now, it's restored to "new" condition, as it was when he first drove it some 63 years ago. It came equipped with factory starter, generator, and battery, but no lights.

Deere was years derelict in diesel power—it didn't have any. Potential customers were buying diesel-powered tractors wearing the competition's colors.

The Model R Diesel wasn't an easy build, and it was field tested rigorously to eliminate any shortcomings. By 1948, Deere was satisfied enough to crank up the production line, and the Model R Diesel began appearing in dealer's showrooms and farmer's fields.

Here, at last, was John Deere's first diesel tractor. It was also Deere's big horse in the stable, with just over 43 drawbar and 48 belt horsepower, according to Nebraska test number 406, conducted in 1949.

The biggest obstacle for any early diesel design was starting, because high compression is synonymous with diesel. A compression ratio of 16:1 and 416 cubic inches put flywheel starting beyond the ability of mere mortals. As recent as the 1950s and early 1960s, starting a diesel engine in zero weather often raised an owner's patience and profanity to an inspired level.

Deere solved the starting problem by building its own gasoline-starting engine for the Model R. The starting engine was built at the Moline Tractor Works, in Moline, Illinois, which had years of experience with small engines thanks to the Model Y, 62, L, and LA.

This opposed two-cylinder starting engine had 23 cubic inches, operated at 4,000 rpm, and generated about 10 horsepower.

The Deere-built, horizontal two-cylinder main engine had a 5.75x8-inch bore and stroke displacing 416 cubic inches and rated at 1,000 rpm. It doesn't seem likely that a diesel tractor of this size would have a thermosiphon cooling system, but it does. The Model R had live PTO, with the hydraulic pump driven by the PTO, so the hydraulics were live as long as the PTO was engaged.

A steel cab was optional. The 5/1 transmission provided forward speeds of approximately 2.12 to 11.5 miles per hour, unless it was rolling on steel, in which case, fifth gear was locked out.

The slanted dash made the gauges easy for the operator to read, but this design lasted only until the 1941 model year when the dash was redesigned to a vertical configuration. This critter is also called a long-hood Model B because the battery is located behind the fuel tank and in front of the dash and steering pedestal.

Deere was about a 15-year Johnny-come-lately to the diesel concept, but when the Model R did arrive, farmers really took notice. The 7,600-pound, standard-tread tractor proved popular with farmers in the Great Plains and the rice country. Consequently, demand and production reached a total 21,293 copies.

Besides being a looker, this 1953 Model 50 has a bevy of notable features: Roll-O-Matic, windowpane rear wheels, and front-wheel weights. It left the factory as a single-front wheel row-crop gasoline model. It was built December 22, 1952, and shipped to Wood River, Nebraska, on December 26, 1952.

# 5 Early 1950s

## From Letters to Numbers

For Deere & Company, the 1950s brought new leadership and new direction to a company that had already seen 120 years of change. Deere's president at the time, Charles Wiman, initiated several of these changes as the decade opened.

Wiman pushed hard for expansion into overseas manufacturing. Although the vote was unanimous, the board's feelings were divided about the new manufacturing facility in East Kilbride, Scotland.

British politics delayed the government from carrying out its part of the agreement. Consequently, Deere's board eventually decided to terminate the project. Although this venture didn't work, Wiman had been successful in focusing the board's attention on the possibility of manufacturing Deere products outside of the United States.

Then, in a rare case of thinking outside the implement box, Wiman convinced the board to jump into the fertilizer business. This time, the board wasn't unanimous, and it wasn't extremely enthusiastic about venturing into a business so different from its forte.

Nonetheless, Deere brought the fertilizer project to fruition, as the Grand River Chemical Division of Deere & Company, in Pryor, Oklahoma. Production began in 1954, but the fertilizer business eventually proved to be too far outside the box, and Deere sold the company to Nipac Inc., of Dallas, Texas, in 1965.

In 1953, it was primarily Wiman who made the decision to discontinue the two-cylinder tractors in favor of four- and six-cylinder models. The top-secret research and engineering for the New Generation of Power was soon underway in earnest. Before the decade closed, the new tooling and production lines for these tractors would need to be in place.

Unfortunately, Wiman died on May 12, 1955, forcing a change in leadership. The burden of bringing these new endeavors to reality fell to Deere & Company's new president, William Hewitt.

### Last of the Letters

Concurrent with these far-reaching activities, the Letter Series tractors—the Model D, Model A, Model B, Model G, Model M, and Model R—were all in production as the 1940s closed and the 1950s opened.

During the 1950s, Deere also introduced three new series of tractors. Because so much was crammed into the 1950s, the decade is broken down into three chapters. This chapter deals with the last Letter Series tractors and the first Numbered Series. Chapter 6 covers the 20 Series tractors, and chapter 7 focuses on the 30 Series tractors.

### Model A

Late in 1949, Phase II, or improved, Model A tractors began production. Major improvements were a single gearshift transmission, two-piece front pedestal, engine cylinder head and block refinements, clamshell fenders, and square rear-axle housings.

Still in their original dress, this 1952 Model A tractor and early 1940s Model 12A combine with a 5-foot header are ready for the harvest field. The tractor was purchased from its original owner in Le Roy, Kansas.

Total production count of the late-styled Model A, in all its different variations, was an impressive 117,965 copies.

### *Model AW*

Approximately 6,275 vehicles designated late-styled Model AW were made, which was a distant second place to the Model A's 106,040 production numbers. The gasoline engine proved twice as popular as the all-fuel version, with 4,314 and 1,961 copies, respectively.

### *Model AO*

Finally, in 1950, the Model AO got the improved 321-ci engine that the other Model A tractor had received in 1947. This engine was available as a gasoline or all-fuel powerplant. Buyers chose the gasoline engine 762 times and the all-fuel version 733 times.

It wasn't until 1950 that the Model AO received the family styling that appeared on the rest of the line in 1947.

The 6/1 tranny featured a creeper first gear, and these tractors showcased Deere's live hydraulic system.

The necessary tinwork to put the AO in full dress was available as an option—engine shields, operator shields, tapered citrus fenders, and fender sides. Or, the buyer could dress it with any combinations of the protection package.

### *Model AR*

Small-grain producers in the Great Plains liked the new standard-tread tractors 9,900 times as a gas burner and 520 times with the all-fuel engine. The styled Model AR was offered from 1950 through 1953.

### *Model AH*

The last variation of the late-styled Model A was a true Hi-Crop vehicle with 33 inches of clearance under the rear axle. The extra height on the rear axle was provided by drop housings that contained the final drives. The extra clearance on the front axle was due to longer spindles, or as some call them, knees.

Mechanically, the Hi-Crop was very similar to the basic Model A Row-Crop. However, there were some necessary changes to the drawbar and three-point hitch to compensate for the higher stance.

The Model AH was available with either a gasoline or all-fuel engine, with 246 tractors produced in the gasoline version and 181 copies made for all fuel. Production began in 1950 and closed in 1952.

## Model B

Late-styled Model B tractors featured an enclosed flywheel, new seat, pressed-steel frame, and electric starter and lights as standard equipment. More user-friendly controls were mounted on the left side of a redesigned steering-column pedestal, and the battery was relocated under the cushioned seat.

Along with the styling improvement came a boost in horsepower due to an increase in the bore to 4.687 inches, while the stroke remained at 5.5 inches. Late-styled Model B tractors displaced 190 cubic inches, and rpm was upped to 1,250.

This immaculately restored Model G Hi-Crop is a 1952 model. It is one of only 235 manufactured by Deere & Company. About 115 of these vehicles were exported, which left only 120 stateside. All Model G Hi-Crop tractors were fitted with the all-fuel engine.

Nebraska tests were conducted on two late-styled Model B tractors in 1947. The gasoline engine's compression was higher than the all-fuel engine due to the use of different pistons.

Nebraska test number 380 was conducted with gasoline as fuel, and it yielded 24 drawbar and 27 belt horsepower.

Test number 381 was conducted with tractor fuel and registered 21 drawbar and 23 belt horsepower. Both test tractors were equipped with the 6/1 transmission.

Production numbers for the late-styled Model B were impressive, with 100,920 units. Gasoline tractors accounted for 89,334 of this total, with all-fuel units numbering 11,586—give or take one or two tractors.

### Model BW

The preference for gasoline at this time was quite obvious and is reflected in the Model B numbers. For the wide-front Model B gasoline tractors, 3,022 were built, while the all-fuel accounted for only 790 copies.

### Model BN

Production figures for the late-styled Model BN were 3,913 gasoline and 608 all-fuel. The Model BN was

Many large-model Hi-Crop tractors went south to the cane fields near the Gulf of Mexico, where the unfriendly conditions turned them into rust buckets beyond salvage. A Hi-Crop vehicle such as this Model G is impressive no matter how you look at it. This unit was built July 15, 1952, and shipped to Tampa, Florida.

designed for vegetable growers who planted in 28-inch or smaller rows. The single front-wheel version was initially advertised as a garden tractor.

## Model GH

Sugar cane farmers in the south liked the extra power of the Model G and the efficiency of burning heavy fuel. In Deere's tradition of customer service, it accommodated these customers by building 235 of the late-styled Model G Hi-Crop from 1950 to 1953.

## Doing It by the Numbers

As the postwar boom gained momentum, agricultural-equipment manufacturers were challenged to provide equipment that could do more and do it faster. The number of tractors on the nation's farms rose from approximately 2,000 in 1910 to 3,609,281 by 1950.

In 1953, after years of head scratching, arguing, and soul searching, Deere & Company reached a decision to replace the beloved two-cylinder engine with new four- and six-cylinder powerplants.

The death knell of the Johnny Popper had sounded. To make this transition without destroying the company was the formidable challenge facing Deere's management team.

It was imperative that the two-cylinder tractor line be kept competitive and profitable while the New Generation multi-cylinder tractors were secretly being engineered and produced to replace the venerated two-lungers. Any hard evidence that the two-cylinder tractors were being discontinued would have killed sales and possibly killed the company.

From the Waterloo Boy of 1918 to the Model R of 1949, it could be argued that the Deere tractor line had evolved primarily from either internal or external crisis—sometimes concurrently. Now, Deere realized it was time to design an entire new line of tractors that would embody all the technological and engineering advances that could be crammed into a farm tractor. Until that became a reality, the Letter Series tractors would have to carry the load.

What emerged off the drawing boards and the assembly lines in 1952 were Deere's new first Numbered Series tractors. In addition to offering more horsepower, the new series raised the bar for operator comfort and convenience. These were features that farmers were beginning to expect, even demand, before parting with their hard-earned dollars.

At the small end of the line was the Model 40, at 18 drawbar horsepower, and anchoring the big slot was the Model 80, which delivered 60 drawbar horsepower.

## Model 40

The Model 40 rolled into the spot previously filled by the Model M and was available in single front wheel, Tricycle, wide front axle, Hi-Crop, and Crawler versions. Approximately 49,000 copies of the 40 Series were built before production closed in 1955.

Deere viewed the series as being totally new tractors and stated, "The Model 40 Series Tractors are new from stem to stern." These tractors did have a long list of truly new features, but a very pronounced shadow of the Model M lurked underneath new tinwork. The 40 Series, like the Model M, was produced at Deere's Dubuque, Iowa, facility.

All Model 40 vehicles used the vertical two-cylinder engine that was vintage Model M, but it was tweaked a trifle. The 4x4-inch bore and stroke remained the same, but the rpm was increased from the M's 1,650 to 1,850, and the compression ratio on the gasoline engine went from 6.0:1 to 6.5:1, while the all-fuel compression ratio was 4.7:1. Of course, the 100 cubic inches remained the same. These features increased the horsepower approximately 15 percent.

The shift from low-grade fuels to gasoline was well underway when the 40 Series debuted, but there was still enough demand for distillates, tractor fuel, and kerosene that an all-fuel engine was offered in all variants of the Model 40. A total of about 1,650 tractors left the Dubuque facility with all-fuel engines.

One of the most important features of the Model 40 Series was the new hydraulic system with industry standard three-point hitch. Like all other tractor manufacturers, Deere finally had to bow to the Ford tractor with Ferguson's revolutionary three-point hitch.

Deere's live Dual Touch-O-Matic control system coupled to the new Load-and-Depth Control feature allowed the operator to independently control either side of the split rockshaft. Load-and-Depth Control regulated the working depth of mounted equipment to ensure that on uneven terrain or in varying soil conditions the planting depth or cultivating depth remained constant.

There were 23 specially designed, rear-mounted three-point implements and four mid-mounted tools for the Model 40 Series.

Optional accessories included an hour meter, full front and rear lighting equipment, belt pulley, wheel weights for front or rear, fenders, radiator shutter, and a cigarette lighter.

A new padded seat, which would eventually evolve into the deluxe seat featured on the New Generation tractors, was one of the most appreciated improvements in operator comfort.

### *Model 40T*

Traditionally, the Tricycle designation was applied to tractors with a narrow front axle with dual wheels or a single front wheel. For the Model 40, the T includes the conventional dual-wheel Tricycle and single front wheel, as well as two wide adjustable front axles.

The Tricycle proved the most popular of the 40 Series machines, with a production total of 17,906 copies. Planting and cultivating specialty crops,

# Timeline: 1950–1955

**1950**

U.S. troops go ashore at Inchon, Korea.

The census puts the U.S. population at 150.6 million. Families living and working on farms account for only 5.4 million.

Industry wages hit a new average weekly high of $60.53.

New in the marketplace: power steering for automobiles, garage door openers, and Dacron suits.

**1951**

Race riot in Chicago prompts Illinois Governor Adlai Stevenson to mobilize National Guard.

The United States detonates the first thermonuclear device, and mankind moves into the H-bomb era.

**1952**

John Deere Model 50: 1952–1956. John Deere Model 60: 1952–1956.

President Truman submits to congress the largest peacetime budget in U.S. history.

General Dwight D. Eisenhower and Richard Nixon are nominated as the Republican's choice for president and vice-president.

Democrats nominate Adlai E. Stevenson and John J. Sparkman.

Republicans win.

**1953**

John Deere Model 40: 1953–1955. John Deere Model 70: 1953–1956.

The Korean War ends.

**1954**

The assembly line at General Motors delivers its 50-millionth automobile.

**1955**

John Deere Model 80 Diesel: 1955–1956.

United States sends military advisers to Vietnam.

especially narrow-row vegetable farming, was the forte of the Model 40T.

Nebraska test number 503, conducted in 1953, verified that the Model 40T with gasoline engine produced 21 drawbar and 24 belt horsepower.

Fitted with the regular axle, a tread width from 48 to 96 inches was possible if the hubs and wheels were reversed. The optional rear axle allowed 48 to 88 inches of rear-tread adjustment. Also, two different front axles were offered, cast iron and tubular steel. Both offered varying tread widths from 48 to 80 inches.

Depending on rear-tire size, the 40T gave approximately 21 inches of crop clearance under the rear axle.

### *Model 40S*

Coming onboard in 1953, the Model 40S was a two-plow General Purpose tractor popular with vegetable growers who planted and cultivated one row at a time.

Rear-wheel tread was adjustable from 38.75 to 55.75 inches, and the front tread was adjustable from 40 to 55 inches. The optional Power Adjust rear wheels could be adjusted for tread width from 39.125 to 66.375 inches.

The Model 40S racked up production numbers of 11,161 gasoline copies and 653 all-fuel tractors, for a total of 11,814 units.

All-fuel engines saved on fuel cost but delivered three less horsepower than the gasoline model, as was reflected in Nebraska test number 504 in 1953, when it registered 21 drawbar and 23 belt horsepower. The tractor-fuel test, number 546, ran in 1955 and garnered only 18 drawbar and 20 belt horsepower.

### *Model 40U*

A Model 40U tractor had to wear several hats. It could be used for general agricultural tasks, or when fitted with an orchard muffler, it served as the Model 40s version of an Orchard. Clad in a coat of industrial yellow or highway orange paint, it became the Industrial Model 40.

A low stance and wide adjustable front axle gave the Model 40U great stability for highway mowing, orchard and vineyard work, or helping on a job site.

Rear tread was adjustable from 40.875 to 57.875 inches and the front-axle tread from 43 to 56 inches. With the optional Power Adjust rear wheel, the maximum tread adjustment was 68.5 inches.

Gasoline models accounted for 5,067 vehicles, while the all-fuel version numbered only 141 of the total 5,208 copies produced.

### Model 40W

The Model 40W Two-Row Utility was produced in 1955. Of the 1,758 copies made, 1,698 were gas burners and 60 were fitted with the all-fuel engine. Production began in January 1955 and ended in October of the same year.

The Model 40W had just 17 inches of clearance under the rear axle but proved to be a popular tractor for cultivating row-crop with the front-mounted cultivator.

### Model 40H

Only 294 copies of the Model 40 Hi-Crop were assembled during production from August 1954 to October 1955. All-fuel versions accounted for 35 units, with the balance gasoline tractors.

Larger rear tires, longer front-axle spindles, and drop housings on the rear axle provided 32 inches of crop clearance. Tread adjustment for both front and rear axles was 54 to 84 inches in 6-inch increments.

To give this high center of gravity machine additional longitudinal stability, the chassis was split and a spacer was inserted between the transmission case

Deere's answer to the Ford-Ferguson 9N was the Model M. This 1950 model has one more horse than the 9N, but didn't sell anywhere near the numbers. However, it posted very respectable production stats—just over 87,800 copies for the M Series.

Call it a Model 40V, a Model 40 Special, or a Model 40 Hi-Clearance, but whatever you call it, it looks nice. This Model 40V first saw the light of day in 1955.

and center frame. For lateral stability, the tread adjustment was restricted to prevent too narrow of a stance.

### *Model 40V*

A Model 40V or Special is a Semi Hi-Crop machine, sometimes called the sweet potato or peanut tractor, that gives 26 inches of clearance, compared to 32 inches for the true Hi-Crop.

Rear-wheel tread was adjustable from 46 to 80 inches, and the front wheels were adjustable from 46 to 66 inches.

This wasn't a high production model; only 329 copies were manufactured—326 as gas burners and only 3 all-fuel tractors.

### *Model 40C*

The Model 40 Crawler was a direct descendant of the Lindeman's design of the BO Crawler and the MC Crawler.

Fitted with the same engine upgrades as the other Model 40s, the 40C produced about 5 percent more horsepower than its predecessor, the MC. The Crawler was tested at Nebraska in 1953, test number 505, at which time the gasoline version yielded 19 drawbar and 24 belt horsepower.

This little Crawler was well received, and production totals reached 11,689 vehicles: 11,407 gasoline units and 282 all-fuel copies.

In July 1953, the original three-roller track was discontinued in favor of either a four- or five-roller design.

Model 40 production was phased out in 1955 to make way for the Model 320.

## Model 50

In 1952, the Model 50 debuted powered by a reincarnated Model B engine. The Model 50 engine retained the 4.687x5.5-inch bore and stroke of the Model B and spun at the same 1,250 rpm. Its first Nebraska test, number 486, gave it 27 drawbar and 30 belt horsepower burning gasoline. However, besides gasoline, buyers could specify either all-fuel or LP-gas engines on their new Model 50.

A year later, according to Nebraska test number 507, the all-fuel version registered 22 drawbar and 24 belt horsepower. In 1955, the LP-gas tractor, test number 540, gave a reading of 28 drawbar and 31 belt horsepower.

Duplex carburetion and Cyclonic Fuel Intake were two new features introduced on the Model 50 and were responsible for producing the extra horsepower. Duplex carburetion wasn't really a separate carburetor for each cylinder. Rather, it was a single carburetor with two barrels. The two-barrel design made it possible to supply identical amounts of the fuel mixture to each cylinder. The Duplex carb wasn't used on the first all-fuel Model 50s, but it was added in October 1953.

The cast arch, or eyebrow, positioned over the intake valve on each cylinder was dubbed the Cyclonic Fuel Intake. This feature improved the air and fuel mixing, resulting in more complete combustion within the cylinder, hence, more power. This feature was also found on the Model 40 engines.

Model 50 tractors offered far more power, operator comfort, and muscle-saving features than the Model B. However, production numbers for the Model 50 didn't come close to the Model B's stellar selling record.

The total number of Model 50s produced adds up to 32,574 units. Of these, 731 were LP-gas tractors, 2,095 all-fuel vehicles, and 29,748 gasoline copies.

## Model 60

Replacing the extremely popular Model A, the Model 60 was the first Numbered Series tractor to be introduced in 1952. Although the total number of Model 60 production doesn't approach the Model A figures, the yearly average was surprisingly close.

The Model A was in production from early 1934 to 1952, resulting in approximately 300,000 copies. During its 18-year run, that's an average of about 16,600 per year. The Model 60 was produced for just over four years, for a total production of about 61,105 units, or a yearly average of 15,276 copies.

### *Model 60N*

First production tractors with the single-front-wheel Model 60 carried the designation N. When the two-piece pedestal was adopted in 1954, this no longer applied, and the N designation was dropped.

### *Model 60W*

The wide front axle Model 60W came packaged as a gasoline, all-fuel with single induction or as an all-fuel with duplex carb. Like the N, the W designation was dropped as soon as the two-piece pedestal arrived.

The PTO-driven belt pulley adds yet another useful feature to this Model 40V at a cost of about $55. However, just to the right of the pulley is a more interesting option. It's a chain sprocket driven from the tractor's right drop axle. This sprocket drove a Dearborn single-row, side-mount planter that was used to plant roses in Louisiana.

The owner said this restoration project quickly progressed from a minor job to a major job. He knew the transmission needed work, but he didn't know that the main case was busted. The end result proves the extra time and money were well spent.

A Model 70's 376-ci two-cylinder diesel puts the horses right here—45 on the drawbar and 50 on the belt/PTO. Deere's promotional material said that's enough for four, 14-inch plow bottoms in any conditions, and five bottoms in some soils.

### *Model 60H*

Introduced at the same time as the Row-Crop Model 60 in 1952, the Model 60H was a true Hi-Crop vehicle, with 32 inches of crop clearance.

Production numbers were 135 copies of the all-fuel version, 62 copies of the gasoline vehicle, and only 15 copies of the LP-gas tractor. Total production for the Model 60 Hi-Crop added up to 212 units.

### *Model 60S*

Five variations of the Model 60 Standard vehicle included: low-seat gasoline, high-seat gasoline, low-seat all-fuel, high-seat all-fuel, high-seat LP-gas.

The numbers read: 1,748 low-seat gasoline, 676 high-seat gasoline, 196 low-seat all-fuel, 209 high-seat all-fuel, and 26 high-seat LP-gas.

### *Model 60 Orchard*

Buyers could choose either a gasoline or all-fuel engine when the Model 60 Orchard tractor was introduced in 1953 for the 1954 model year. The LP-gas engine was made available in late 1956. Production ended in early 1957.

The total number of Model 60 Orchard units was 872 divided between three fuel choices—297 gasoline, 530 all-fuel, and 45 LP-gas.

## Model 70

Considered the replacement for the Model G, the Model 70 offered a significant horsepower boost over its predecessor's 34 drawbar and 38 belt rating. It also offered all the advanced features of the Model 40, 50, and 60. A diesel engine became available in 1954, making it Deere's first diesel-powered row-crop tractor.

The Model 70 engine wasn't a one-fits-all design, raher each engine was designed for a specific fuel type. The all-fuel tractor used the same 413-ci engine of the Model G, with 6.125x7-inch bore and stroke. Engineers coaxed extra horsepower out of the Model 70's 413 cubic inches by increasing the compression ratio to 4.6:1 and adding duplex carburetion.

When tested at Nebraska in 1953, test number 506, these improvements registered 39 drawbar and 43 belt horsepower from the all-fuel engine.

Model 70 tractors came in three variations: the Model 70T Row-Crop, Model 70S Standard, and Model 70H Hi-Crop. All were available with either gasoline, all-fuel, LP-gas, or diesel powerplants.

Production numbers were 35,868 Row-Crop, 5,004 Standard, and 157 Hi-Crop tractors, for a grand total of 41,029 Model 70 copies.

This 1956 Model 80 diesel is one of only 220 built without power steering and consequently was issued with a larger steering wheel than the models with power steering.

## Model 80

To borrow a line from Henry Ford, buyers could have the Model 80 anyway they wanted it as long as it was equipped with a standard front axle and a diesel engine. It was the last of the first Numbered Series to be introduced, and only about 3,500 were built during its short run.

Deviations from the Model R, which it replaced, included a bore increase from 5.75 to 6.125 inches, which boosted the displacement to 471 cubic inches. This was coupled to a 125 rpm increase, to 1,125. It featured a six-speed transmission and a V-4 gasoline-starting engine.

This diesel unit replaced the Model R as the most powerful tractor in Deere's lineup. In 1955, Nebraska test number 567 pegged the Model 80's horsepower at 60 drawbar and 65 belt.

The dash, with full instrumentation, on a 1956 Model 80 Diesel.

Slant-steer Model 320 Standard serial numbers begin at 325001. This 1957 model, serial number 325045, is the 45th such tractor manufactured. It's one of 317 slant-steer Model 320 Standards built with the gasoline engine.

# 6 Mid-1950s
## Bigger and Better Numbers

When Deere & Company President Charles Deere Wiman died on May 12, 1955, William Hewitt, Wiman's son-in-law, was elected as the sixth president of Deere & Company. Hewitt had been at his new post only a few days when he announced, "We are aiming to be first in all our business activities."

Deere, which at this time was still second to International Harvester, would soon become number one in the agricultural machinery industry if Hewitt had his way.

John Deere 20 and 30 Series tractors played an extremely important role in achieving this goal and in convincing customers that the two-cylinder was only going to get better as time went along.

### Model 320

Built in Dubuque, Iowa, the Model 320 tractors offered many mechanical up-grades and improvements while still filling the low purchase cost and low production cost slot in Deere's agricultural and Industrial line. The Model 320's family lineage goes back mainly to the Model M and, to a lesser degree, to the Model 40.

Many of the features of its bigger siblings were incorporated, such as a new three-point hitch with Load-and-Depth Control, Touch-O-Matic hydraulics, adjustable coil-spring cushion seat, and independent disc-type brakes.

The vertical two-cylinder Deere-built engine had 4x4-inch bore and stroke yielding 101 cubic inches of displacement and rating at 1,650 rpm. The 4/1 transmission gave forward speeds of 1.625 to 12 miles per hour.

The cooling system was simple thermosiphon, although Deere advertising material spelled it Thermo-Siphon, which suggests perhaps marketing intended to present it as a Deere feature, like Powr-Trol and Touch-O-Matic.

The timeline for the 20 Series tractors began with the Dubuque-built 420 Series introduced in November 1955. In August 1956, the Waterloo-built 520, 620, 720, and 820 Models were introduced along with the Dubuque-built Model 320 Series.

Styling wasn't a radical departure from the Model M or Model 40, however. The yellow band added to the hood and radiator cowling made them appear as a new line of tractors. This new look helped convince customers, and the competition, that Deere was committed to the continuance of the two-cylinder tradition.

With only 3,084 copies of the Model 320 produced, it might appear that sales were not good. It is important to remember that during this time period, the number of American farms was declining while the amount of work a single tractor could perform was increasing. Also, the switch from horse and mule to tractors for farm power had virtually been achieved by this time. So, the bottom line is that there were fewer potential tractor customers, and the industry would never again see the heady years with production numbers in the hundreds of thousand of a single model.

A 1956 Model 320 Utility made a handy haying tractor on any size farm. Fitted with the orchard muffler, as this one is, the low profile lets it work in places with low overhead, such as dairy loafing sheds. The vehicle is a whisker over 50 inches at the hood.

Just how many horses did the Model 320 have? Officially, we don't know, since it was never tested at Nebraska. Deere rated it at 21.5 belt horsepower.

The 320 Series fielded only four variants—the Standard, Utility, Industrial, and the Special, or V, model, and they were available with either gasoline or all-fuel engines. One other distinctive design feature that divides the series is the straight-steer and slant-steer versions, which refer to the steering wheel's position: either straight vertical or slanted with the top forward.

### *Model 320 Standard*

The Standard version was the sales leader among the Model 320 tractors, with 2,167 copies sold. The breakdown of variants includes 1,836 straight-steer gas, 12 straight-steer all-fuel, 317 slant-steer gas, and 2 slant-steer all-fuel.

Standard equipment included a three-point hitch system, Load-and-Depth Control, electric starter, battery, generator, adjustable front axle, adjustable cushioned seat, adjustable backrest, transmission-driven PTO, fenders, water temperature gauge, oil pressure gauge, fuel filter, oil filter, and air cleaner.

As with its predecessor, the Model M, the Model 320 was equipped with an automotive-type foot clutch and disk-type foot operated brakes. The front axle was adjustable from 40 to 55 inches, while the rear tread could be adjusted from 38.75 to 54.25 inches.

With the final-drive assemblies positioned downward and taller front axle spindles, the Standard model mustered 21 inches of crop clearance.

### *Model 320 Utility*

The Model 320 Utility's low-slung stance provided only 11 inches of axle clearance, however, its low profile—only 50 inches at the hood line—made it an excellent tractor for field, orchard, and vineyard applications.

Today, this nicely restored 320 Utility is wearing agricultural green. The accompanying image of the ID tag shows that sometime in the past, it spent part of its career painted industrial, or highway, orange.

Numbers for the Utility version are 716 straight-steer gasoline, 2 straight-steer all-fuel, and 199 slant-steer gasoline, for a total of 917 units. No slant-steer all-fuel copies were produced.

Utility vehicles carried the same specs as the Standard, with a couple notable exceptions. The Utility tractor could be ordered minus the three-point hitch system if that was the purchaser's preference, and Dual Touch-O-Matic wasn't an option, as it was on the Standard model.

### *Model 320V*

An estimated 60 to 70 Model 320 Standard tractors were converted to Southern Specials or V vehicles. These special high-clearance models were the result of marrying 10-34 rear tires, fenders, drawbar, hitch, and front axles from the 420 V to the Model 320 Standard. The added crop clearance made it desirable for specialty bedded crop cultivation.

### *Model 320I*

A low center of gravity, a coat of yellow or orange paint, and a mounted mower transformed the Model 320 into a popular tractor for maintaining state and county highway right of ways. These Industrial tractors' serial numbers fall within the Utility model numbers, so the exact number produced isn't known.

## Model 420

The Dubuque, Iowa-built Model 420 Series debuted in November 1955, and was the first of the 20 Series to be introduced. They were followed eight months later by the rest of the 20 Series line.

Production can be divided into three distinct groups or phases: Phase I included tractors that were almost dead ringers for their predecessors, the Model 40s. All green tinwork, yellow wheels, and a chrome-plated John Deere medallion on the upper grille served as identification points for the early 420 Series.

Museum quality is obvious when you look at this slant-steer 1958 Model 420 with only 1,600 original hours. But, it wasn't a museum piece when the current owner spotted it in a shelterbelt west of Clearwater, Nebraska. The tractor had been parked there with the exhaust uncovered, and the engine was full of water. The owner said he'd trade the Model 420 for an overhaul job on his Model B. Now, both tractors are in good shape and both owners are happy.

Phase I saw the production of seven different versions of the Model 420 including Standard (S), Utility (U), Tricycle (T), Row-Crop Utility (W), Hi-Crop (H), Special (V), and Crawler (C).

Phase II tractors carried the yellow vertical and horizontal stripe on the hood, and the chrome John Deere was replaced with the leaping Deere badge. These small changes provided the desired new, modern look. The Phase II line added the industrial version Special Utility (I) to the original seven models.

With the exception of the Hi-Crop and LP-gas tractors, a slant steering wheel and automotive dash set the Phase III vehicles apart. Power steering was offered as an option on all wheel models beginning with Phase III production.

All versions of the Model 420 were powered by a Deere-built vertical two-cylinder I-head engine with 4.25x4-inch bore and stroke that yielded 113 cubic inches. This was an increase of 0.25-inch in the bore, which added 12 cubic inches of displacement over the Model 40 powerplant. The rpm remained at 1,850.

Increased compression ratios, combustion chamber changes for better mixture and combustion, plus a new carburetor were additional under the hood modifications and improvements.

The thermosiphon cooling system wasn't adequate for the 420 Series' higher horsepower, so the change was made to a water pump and a pressurized system.

"Handy live Touch-O-Matic hydraulic system and Load-and-Depth Control are built-in features of all 420 wheel-type tractors, as is the convenient, time-saving standard 3-point hitch," were features that the company brochure highlighted.

The Dubuque factory produced many custom orders for buyers. What was standard equipment and what was optional equipment can be unclear. The following is a list of options that could be had on most models sometime during production of the 420: All-fuel engine, LP-gas engine, live PTO, five-speed transmission, fenders, belt pulley, power steering, directional reverser, orchard muffler, silencer muffler, remote hydraulic control, remote cylinder, front- and rear-wheel weights, Power Adjust rear wheels, and hour meter.

Three different versions of the 420 were put through their paces at the Nebraska test lab in 1956. Test number 599 was a 420S burning gasoline, with results of 26 drawbar and 28 belt horsepower.

Test number 600 was a 420S with its engine set up to burn tractor fuel. The horsepower yield was predictably lower, at 21 drawbar and 22 belt. Test number 601 was a 420 Crawler running on gasoline, with results of 23 drawbar and 28 belt horsepower.

Deere & Company made sure that regardless of which Model 420 the customer purchased, there was a matching Deere implement designed and manufactured to handle any farming operation. The list included at least three-dozen implements and attachments, from plows to planters, from peanut pullers to cotton pickers, and from corn shellers to cultivators.

A no-cost option for all Crawlers and Utility models was an industrial yellow paint job. In fact, the same yellow paint was available as a special order for any of the 420 tractors.

The current owner of this 1957 slant-steer Model 420 found the tractor on a ranch in Rocky Ford, Colorado. The rancher didn't want to sell but finally agreed to trade for a Model 50 Row-Crop unit. The 420 has the five-speed transmission and live PTO. Options include three-point hitch, swinging drawbar, and speed-hour meter, but not power steering.

A 1958 Model 620 in original dress. It's a gasoline engine Row-Crop vehicle with adjustable front axle. Of the 21,069 copies of the 620 Row-Crop tractors, it's doubtful that any left the factory with fewer options than this one, serial number 6221395.

The 420 models were good sellers, with an cumulative total of 46,450 vehicles rolling off the Dubuque assembly lines.

### *Model 420C*

The demand for track-laying machines to work steep hillsides, boggy fields, orchards, groves, and industrial sites resulted in more sales for the 420C than any of its wheeled siblings. Its total production reached 17,882 machines, including both the four- and five-roller versions. There were 234 all-fuel and only 4 LP-gas Crawlers produced.

Track shoe width could be 10, 12, or 14 inches, with regular tread width from 38 to 44 inches. Beginning with Phase II production, a three-point hitch was made available as an option. This same hitch could be retrofitted to earlier Model 420C and even the Model 40C.

### *Model 420W*

The Model 420W racked up more numbers than any other wheeled Model 420, with a total of 11,197 vehicles manufactured. There were 399 all-fuel copies of the 420W and 100 LP-gas tractors produced.

Standard transmission for the 420W was a 4/1, with a 5/1 offered under options. The 5/1 added a 6.25 miles per hour gear to provide a higher travel speed for light fieldwork.

Fitted with the regular front axle, tread width was adjustable from 48 to 80 inches. The optional front axle provided a tread width from 56 to 88 inches.

Regular rear-tread width could be adjusted from 48 to 98 inches, or if the owner had opted for the Power Adjust rear wheels, the setting ranged from 56 to 88 inches. It stood taller than the Utility model and offered 21 inches of crop clearance.

With the ability to cultivate up to four rows, perform any job on small acreage farms, plus do its fair share on big operations, it's no surprise sales numbers were good.

### Model 420T

Second-highest production numbers among the Model 420 wheel tractors belong to the Tricycle version, which posted a total of 7,580 copies. Its popularity and versatility were due in part to four interchangeable front ends, which included the dual front-wheel Tricycle, the single front wheel, and two wide adjustable front axles. The square tube design front axle allowed tread width adjustment from 48 to 80 inches, while the round tube design adjusted from 68 to 88 inches.

Equipped with 28-inch rear wheels, the 420T provided 21 inches of crop clearance. Standard equipment included three-point hitch and dual Touch-O-Matic.

All-fuel production totaled 234 units, and the 420T was the leader in LP-gas vehicles, with 225 copies made.

### Model 420U

"The Handy, Economical, Low-Built 2–3 Plow Tractor for Field, Orchard, Grove, and Vineyard Work" is how John Deere described the 420 Utility tractor. With a no extra cost optional coat of yellow paint, it also masqueraded as an Industrial and could be found working on construction sites or maintaining road right of ways.

When production ceased, 4,932 Model 420U tractors had made the journey down the Dubuque assembly line. Only 6 of those were LP-gas units, while 54 were fitted with the all-fuel engine.

If the customer didn't need or want the standard three-point hitch, he could order his tractor without it. The Utility version was the only Phase I Model 420 to offer power steering as an option from the very beginning of production.

A wide adjustable front axle was standard, but with only 11 to 14.50 inches of crop clearance. it wasn't designed with row-crop in mind.

### Model 420S

Considered a one-row tractor with 2–3 plow power, the Standard version of the Model 420 added yet another choice for the 420 Series customer.

Total production of the Standard model reached 3,908 during its lifetime, with 69 of those outfitted with the all-fuel engine and 23 fitted for LP-gas fuel. Although dual Touch-O-Matic wasn't regular equipment on the 420S, it was offered as an option.

### Model 420H

The Model 420H proved to be one of the most popular Hi-Crop tractors John Deere ever manufactured, with 610 copies produced. With the exception of 47 all-fuel tractors and 4 LP-gas vehicles, all were gasoline burners. Crop clearance underneath this tractor measured 32 inches, making it ideal for bedded crops.

To add longitudinal stability, the 420H continued the practice of adding an approximately 7-inch spacer in front of the transmission, as was introduced on the Model 40 Hi-Crop.

The customer who special ordered this Model 620 liked his tractors with no frills and didn't want power steering or a speed-hour meter. Just gas it up, hook it up, and go work the dirt.

# Timeline: 1956–1958

**1956**

John Deere Model 420: 1956–1958.

Congress passes farm subsidy bill, paying farmers to reduce crop production.

Nikita S. Khrushchev makes his famous statement, "Whether you like it or not, history is on our side. We will bury you."

President Eisenhower wins second term by a landslide vote.

**1957**

John Deere Model 320: 1957–1958. John Deere Model 520: 1957–1958. John Deere Model 620: 1957–1958. John Deere Model 720: 1957–1958. John Deere Model 820: 1957–1958.

American Motors' Rambler is the industry's first compact automobile.

The 707, the first American jet transport, carries 219 passengers at a cruising speed of 600 miles per hour.

America's answer to the Soviet's Sputnik, the Navy Vanguard rocket with a 3-1/4-pound satellite payload, gets only 2 feet off the launch pad before exploding.

**1958**

Pan American World Airways begins the first American jet service to Europe.

Three-point hitch was an option, while dual Touch-O-Matic was standard equipment.

The Hi-Crop never received the slant steering wheel during Phase III production, and the Directional Reverser and Power Adjust rear wheels were never available on the Hi-Crop model.

### *Model 420I*

Designed off the Row-Crop Utility, the Special Utility was fitted with stronger, fixed-tread front axle and heavy-duty steering arms to take the punishment of front-end loaders used day-after-day on commercial job sites. To benefit the operator, this tractor offered power-assisted steering as an option.

There was no Phase I production of the Special Utility model, because it didn't join the line until April 1957. Only 255 copies were produced during Phase II and III, all of which were fitted with gasoline engines.

### *Model 420V*

Semi Hi-Crop and Southern Special were other names for the Model 420V. It wasn't a true Hi-Crop, but it did provide 26 inches of under-the-axle clearance. It didn't have the spacer in front of the transmission like the true Hi-Crop, and it used 34-inch rear tires and a shorter adjustable front axle. Like the Hi-Crop, it retained the straight vertical steering wheel throughout its production. Standard equipment included a three-point hitch and dual Touch-O-Matic.

The V Special had the lowest total production numbers of any Model 420 version, with only 86 copies produced. All were gas burners, with the exception of three all-fuel tractors.

### Model 520

The Model 520 was in the tradition of the Model B and Model 50 but was in fact almost 100 percent new tractor. The engine had the same 190 cubic inches from the same 4.687x5.5-inch bore and stroke, but that's where the similarities ended.

Improved combustion chamber design, higher compression ratio (gasoline and LP-gas), and a boost in rpm jumped the horsepower output to 33 drawbar on gasoline compared to 27 drawbar for the Model

50, and the Model B could only muster 16 drawbar horses. By 1947, the Model B had only climbed the horsepower ladder to 24 drawbar as a gas burner. The enormously popular Model B was definitely evolving upwards in the guise of the Model 520.

Three fuel options were offered for the 520: all-fuel, gasoline, and LP-gas. The new engine speed was 1,325 rpm, up from the previous 1,250 of the old Model 50 engine. Engineers continued to squeeze more horses out of every cubic inch. The rest of the tractor was getting better, too. The 6/1 tranny was beefed up, as were the final drives to handle the increase in engine power.

The new tractors were tested three times at Nebraska in 1956, with the gasoline and LP-gas engines posting the same results: 33 drawbar and 37 belt horsepower in test numbers 597 and 590, respectively.

The best the all-fuel, or tractor fuel, model could muster was 23 drawbar and 25 belt horsepower, according to test number 592.

The Model 520 was a row-crop tractor, period, with no Standard, Hi-Crop, Hi-Clearance, or Crawler produced. But, there were a lot of options in the row-crop front-end equipment: Dual-tricycle without Roll-O-Matic was standard equipment. Options included dual-tricycle with Roll-O-Matic, single wheel, adjustable wide axle from 48 to 80 inches, and 38-inch fixed tread.

A few changes were made for the 1958 model year, including the axle-mounted step, which became standard equipment. The dash panel was painted black. Sealed-beam lights were adopted, and the steering wheel received a new plastic covering. Deere & Company advertised the Model 520 as a three-plow tractor capable of handling four-row planting and cultivating equipment.

A total of 13,048 Model 520s were manufactured at Deere's Waterloo, Iowa, plant.

## Model 620

The extremely popular Model A and the Model 60 comprised the Model 620's family tree. The 620 offered a few more horses under the hood and was a four-plow tractor that approached the half-century horsepower mark on the belt.

The same family traits of the 520 applied to the Model 620 regarding standard equipment and optional equipment. However, the Model 620 was offered in Standard, Hi-Crop, and Orchard versions. Row-crop vehicles accounted for the lion's share of production, with 21,069 copies. A distant second place was the Standard version, with 988 produced; followed by the Orchard model, with 427; and the Hi-Crop tallied only 48 copies.

The Nebraska test lab was busy with John Deere 20 Series tractors in 1956. Like the 520, three different versions of the 620 were tested that year. Test number 598 was conducted on gasoline, number 604 on tractor fuel, and number 591 on LP-gas. Horsepower results were 42 drawbar and 46 belt for gasoline, 32 drawbar and 34 belt for tractor fuel, and 44 drawbar and 49 belt for LP-gas.

### *Model 620T*

Row-Crop designation included the two-wheel Tricycle, the two-wheel Tricycle with Roll-O-Matic, single front wheel, adjustable front axle from 48 to 80 inches, and the 38-inch fixed-tread front axle.

Sales for the Row-Crop were respect-able, with a total of 21,069 vehicles produced during its two year run.

### *Model 620S*

The Model 620 Standard was primarily sold in the wheat-growing Great Plains and the rice country of Arkansas, Louisiana, and Texas. Only 988 copies were sold. Gasoline models accounted for 920 of the sales, LP-gas 37, and all-fuel 31.

### *Model 620H*

An LP-gas Hi-Crop Model 620 comes close to the "very rare" category, since only eight copies were produced. Sixteen all-fuel models were manufactured, and 24 units of the gasoline version were made. The total of 48 makes the 620H a low-production tractor, regardless of the fuel option.

It's one of only 37 ever manufactured, and, according to company records, this Model 620 LP gas tractor was shipped to Tampa, Florida, on December 12, 1956. It eventually found its way to La Crosse, Kansas, and the present owner bought it at auction west of Hutchinson, Kansas. Its serial number is 6205652.

### *Model 620 Orchard*

The Orchard version was the last Model 620 to join the lineup, but it lasted the longest—1957 to 1960. Rather than tool up to offer a Model 630 Orchard version, it seems the decision was made to carry over the 620 Orchard tractor to fill that slot in the 30 Series lineup. Total sales amounted to 427 copies.

## Model 720

Deere's Big Daddy Row-Crop was available in four fuel options: diesel, LP-gas, gasoline, and tractor fuel (all-fuel engine). This Model G's offspring was Deere's replacement for the Model 70, and total production of the Model 720 tallied 27,573 vehicles.

The evolution of higher rpm engines continued with the gasoline, all-fuel, and LP-gas powerplant for the 720 tractors. The bore was increased from 5.875 inches to 6 inches, and the stroke was shortened from 7 inches to 6.375 inches. This dropped the cubic inch displacement from 379 to 360, but those 19 cubic inches weren't missed, as improvements in the piston, cylinder head, and ignition system more than made up the difference in horsepower.

The rpm on the gasoline, all-fuel, and LP-gas engines was kicked up to 1,125 from 975 rpm on the Model 70.

The diesel engine retained the same specs as the Model 70, with a 6.125 x 6.375-inch bore and stroke providing 376 cubic inches. The rpm also stayed the same, at 1,125. The Model 720 Diesel was introduced with a gasoline-starting engine. Later, an electric-start package was available.

When the present owner purchased this tractor, it didn't look showroom-new like it does now. It was missing the LP-gas tank and the engine block, and most of the tractor was in various stages of disassembly. The owner had a 620 LP-gas Row-Crop unit that he used as a donor tractor.

In 1956, Nebraska test number 593, using LP-gas, recorded 52 drawbar and 57 belt horsepower. Test number 594 was with diesel fuel, with results of 51 drawbar and 56 belt horsepower. Test number 605 was with gasoline, and yielded 53 drawbar and 57 belt horsepower. Tractor fuel was the fuel for test number 606, with the results reading 40 drawbar and 44 belt horsepower.

### *Model 720T*

Basically, the same front-axle options as offered for the other 20 Series Row-Crops were available on the Model 720 Row-Crop. Customers liked the power and convenience of this tractor and purchased 22,950 copies.

### *Model 720S*

The Standard 720 offered buyers a choice of two front-axle offerings—the 55.5-inch fixed tread or the adjustable 52–68-inch tread. Approximately 4,523 units were produced: 520 gasoline, 80 all-fuel, 345 LP-gas, and 3,578 diesel. Many of the all-fuel Standards were exported.

### *Model 720H*

Collectors go to the ends of the Earth to find a Model 730 Hi-Crop, and they are certainly worth the effort. The build numbers for the Model 720 Hi-Crop and the Model 730 Hi-Crop are almost identical—125 to 123, respectively. Mechanically, they are almost identical. As far as collecting goes, the Model 720 Hi-Crop may be a sleeper.

## Model 820 Diesel

The Model 820 follows in the footsteps of the Model R and Model 80 as Deere's big Standard tractor. The Model 820 weighed in at 8,729 pounds and churned

This 620 Standard LP-gas tractor is loaded and has almost all the options, including PTO, three-point hitch, and Float-Ride seat.

out 67 drawbar and 72 belt horsepower at 1,125 rated rpm, as verified by Nebraska test number 632.

Buyers had one fuel option—diesel. The Standard model came in two variations, Wheatland or Rice Special.

About the only external changes that distinguished the first production Model 820 from the Model 80 was the yellow paint stripe on the hood, the hood medallion, and the new fenders.

Under the hood, the first production 820 tractors utilized the same powerplant as the Model 80. However, Deere engineers upgraded a few items, including the connecting rods, the brakes, and the power steering.

To ensure faster starts in extremely cold weather, a V-4 starting engine was part of the package. Production models were never available with electric start.

Oversized or elongated clam-shell, sometimes called elephant-ear, fenders provided a cleaner platform and, therefore, a cleaner and happier operator.

The option list included an All-Weather steel cab, Float-Ride seat, radiator shutter, cigarette lighter, foot-operated throttle, creeper gear (1.65 or 2.35 miles per hour), weather brake, custom Powr-Trol, Power Steering, live PTO, remote hydraulic cylinder(s), and electric oil and coolant heaters. Steel wheels were available as a special order item.

Official 820 Industrial vehicles don't exist, but an agricultural unit could be ordered with Industrial-tractor options, including a yellow paint job.

Approximately 6,859 Model 820s were produced. About 3,018 were green-dash tractors, and 3,840 were black-dash machines.

About 460 customers declined the power steering option.

The dash on this Model 620 LP-gas tractor has only one unfilled knockout, and that's for the cigarette lighter.

If you tally all the agricultural Model 830s produced, the total is approximately 6,585 copies. Of those, 3,076 had electric start, like this 1960 Rice Special.

# 7

# Late 1950s

## Closing Numbers, Final Curtain

Deere & Company's competition called the 30 Series tractors obsolete, but hindsight suggests that "brilliant" would best describe these final two-cylinder tractors.

Teaming up with industrial designer Henry Dreyfuss, Deere engineers produced a stunning new John Deere tractor line. The new look included a more rounded hood line, a slanted dash and steering wheel, and a slightly different yellow hood stripe. The model-number typeface style was changed and moved from low to high on the side of the radiator cowl.

The dash also received a new look and arrangement. Its forward slant, along with the slanted steering wheel, changed the tractor's side profile dramatically and placed the gauges in an easy-to-read position.

Underneath the new appearance, the 30 Series tractors were virtually identical to their 20 Series predecessors. The significant difference necessary to herald them as the "New John Deere" tractors lay primarily in appearance, comfort, and convenience. Mechanically, there just wasn't much difference.

Deere & Company hoped these new cosmetic features would remove any customer's concerns that it was about to abandon the trusty two-cylinder. Marketing loved the new series, because even though they were selling the same tractor, they now had some sizzle to sell.

Many of these small design changes were really the Dreyfuss team experimenting with, and perfecting, what would become features on the New Generation tractors.

The new styling was effective but extremely conservative compared to automobile designs of the period. Fins were in and American's taste for big, bold, road cruisers was unrestrained. Dreyfuss, however, practiced some restraint in designing the farm tractor and focused on what he called the human factor—ease of operation and operator comfort.

Leading the way in the comfort and convenience category was a seat that set new standards for the industry. This advanced seat design was carried over to the New Generation tractors, so owners who purchased a 30 Series tractor were literally sitting on the future.

A long list of options was offered: speed-hour meter, fuel gauge, cigarette lighter, air stack and precleaner, clam-shell fender, styled flat-topped fender with lights, weather brakes, Float-Ride seat, rear muffler, custom Powr-Trol, single or dual remote hydraulic cylinder equipment, single or dual front-mounted Rockshaft, universal three-point hitch, Load-and-Depth control, live PTO, front-frame weights, front-wheel weights, and rear-wheel weights.

With the exception of the Model 435, the 30 Series tractors were introduced in the fall of 1958 for the 1959 model year. They stayed in the line until the New Generation tractors were introduced in the fall of 1960.

### Model 330

While the Model 320 machines could be purchased with an all-fuel engine, the Model 330 tractors were available only as gas burners.

This is a 1959 Model 330 Utility. It's one of 247 produced and has been restored to mint condition.

There were four variants of the Model 330: the Standard, Utility, Industrial, and the Special.

Standard model production accounted for 844 copies, and the Utility model was a relatively low-production vehicle, with only 247 copies made. This brings the total number of Model 330s produced to 1,091 units.

The Model 330V, or Special, was not an official Deere & Company product, but a few have turned up. It's probable that 330 Standard tractors were converted with appropriate parts from a Model 40V or a Model 420V.

## Model 430

The two-cylinder engines were the same as featured in the 20 Series tractors. No 430 Series tractors were tested at Nebraska, but it certified that the 430 tractors had the same test results as the 420 tractors. See the test results for Model 420 tractors or refer to Nebraska test numbers 599, 600, and 601.

Model 430 tractors were available in three fuel options: gasoline, all-fuel, and LP-gas. All 430 vehicles, regardless of which model, received sequentially numbered serial numbers, starting at 140001 and ending with 161096.

### *Model 430S*

Anyone seeking a low-production 30 Series tractor should consider the LP-gas 430 Standard, as only five were produced. Eighteen all-fuel versions were manufactured, along with 1,786 gasoline copies.

### *Model 430U*

Deere's Dubuque, Iowa, facility produced 1,340 copies of the Model 430 Utility. Three of these were LP-gas models and 10 were all-fuel versions. By then, the customer's choice of gasoline accounted for the vast majority of tractors built.

### *Model 430W*

The Model 430W (Row-Crop Utility) accounted for 5,981 units—about twice as many as any other version of the Model 430 tractors. Of these, 68 were LP-gas, 88 were all-fuel vehicles, and the rest were gasoline.

A 1960 Model 330 with optional front frame weights and rear-wheel weights. The inner front weight is listed at 144 pounds, the outer front at 147 pounds, and the rear-wheel weights at 148 pounds. The collector who restored this tractor took the unit completely apart and replaced or rebuilt everything.

### *Model 430T*

The Tricycle was the second most popular 430 model, with 3,255 copies manufactured. It was also the most popular LP-gas tractor of the 430s, with 128 copies produced. There were 33 all-fuel vehicles made in the Tricycle configuration.

### *Model 430H*

The 430H's dash and steering wheel received the same styling as the rest of the line. An exception was the five custom LP-gas Hi-Crop 430s built at the Dubuque facility. These were a throwback to the Model 420, with vertical steering wheel and dash.

Just 27 all-fuel copies of this Hi-Crop tractor were produced, making it a very desirable addition to any collection. Adding the 183 gasoline vehicles produced brings the total number of 430 Hi-Crop tractors to 215.

### *Model 430V*

All 63 copies of the Model 430V, Special, Southern Special, or Semi Hi-Crop tractors were gasoline burners. This was a high clearance version of the standard tractor and provided 26 inches of clearance under the axle.

### *Model 430C*

Crawler customers buying a Model 430C could specify either a four- or five-roller frame, and they had the choice of a gasoline, all-fuel, or LP-gas engine. Customers chose the all-fuel option 33 times and the LP-gas engine only four times. Gasoline units numbered 2,203, making a grand total of 2,240 copies of the 430 Crawler.

Both a yellow Crawler and yellow-wheel version of the Model 430 were offered through Deere's Industrial Division.

Because the 430 tractors carried the same engine and drivetrain as the 420 models, the Nebraska test lab certified that the 430 tractors had the same test results as the 420 tractors.

A 1960 Model 430T Row-Crop unit with the tricycle front end. This is just a real straight Model 430 with five-speed transmission and three-point hitch. It's another tractor that was taken completely down and rebuilt by a 30 Series collector.

## Model 530

Tracing the Model 530's lineage, you'll find the Model 520, the Model 50, and the Model B. Although the Model 530 filled an important slot in the Deere tractor line, it didn't begin to approach the Model B's popularity or production numbers, with only 9,763 copies made. Unlike the 330 and 430, the Model 530 is a Waterloo-built tractor.

Concerning major components, mechanically, the Model 530 was a dead ringer for the black-dash Model 520. Style and operator conveniences received what time and budget the engineering department had available for the new Model 530.

The Model 530 was a General Purpose Row-Crop vehicle only, with the dual-tricycle front end as standard equipment. Other front-end options included the dual-tricycle with Roll-O-Matic in regular or heavy-duty versions; a regular or heavy-duty single front wheel; an adjustable wide front axle; and the 38-inch fixed-tread front axle.

# Timeline: 1959–1960

**1959**

John Deere Model 330: 1959–1960.
John Deere Model 430: 1959–1960.
John Deere Model 530: 1959-1960.
John Deere Model 630: 1959-1960.
John Deere Model 730: 1959–1961.
John Deere Model 830 Diesel: 1959–1960.
John Deere Model 435: 1959–1960.

Fidel Castro visits the United States after Cuban Revolution brings him to power in Cuba.

Ford scraps the Edsel after fewer than 100,000 are sold in two years.Alaska and Hawaii become 49th and 50th states, espectively.

**1960**

Census places U.S. population just short of 180 million.

Democrats nominate John F. Kennedy for president, while the Republicans trot out Richard M. Nixon. Kennedy wins.

Deere & Company's New Generation tractors stun the industry when introduced in Dallas, Texas.

This is the owner's story on this interesting 1959 Model 630. "That tractor didn't come on steel, but it is a factory option for the 30 Series tractors that's shown in the parts book. I don't believe the extensions are shown in the parts book, but they are correct. I found this new old-stock that had never been on a tractor, and I thought it would be neat to convert that tractor to steel. This makes only the third 30 Series tractor that I've ever seen on steel. It's very rare."

The agricultural Model 530 was sold, or at least offered, by the Industrial Division, but there was never a production Industrial model. Perhaps a few were painted yellow and masqueraded as 530 Industrials.

Once again, because the 530 tractors carried the same engine and drivetrain as the 520 models, the Nebraska test lab certified that the 530 tractors had the same test results as the 520 tractors. See the specs for the Model 520 or refer to Nebraska tests number 590, 592, and 597.

## Model 630

The pedigree of this tractor reads from the Model A to the Model 60 to the Model 620 to the Model 630. The story of the Model 630 is almost a carbon copy of the Model 530, with some notable exceptions.

Besides the Row-Crop, the Model 630 was available as Standard and Hi-Crop vehicles. Excluding tractors made in Mexico, total production of all versions of the Model 630 tallied 18,057.

Row-Crops accounted for the vast majority, with 15,281 gasoline, 1,878 LP-gas, and 216 all-fuel tractors produced. Standard production was 706 gasoline, 16 LP-gas, and 25 all-fuel vehicles.

The owner of this tractor did custom corn picking and was in the Herington, Kansas, John Deere dealership looking at a John Deere 55 combine and two-row corn head. The dealer told him if he bought the outfit, he could use it for the harvest season and then trade it back for other new machinery of the same dollar amount. After harvest, he took the combine back and went home with this 1960 Model 530 with a quick-tooth cultivator, a grain drill, a planter, and a spring-tooth harrow.

Hi-Crop production definitely falls in the very rare category, with 11 gasoline copies, 3 LP-gas copies, and 5 all-fuel copies.

As with the previous 30 Series tractors, the Model 630s weren't tested. Nebraska did, however, certify that the 630 tractors had the same test results as the 620 tractors in test 591, 598, and 604.

This 1960 Model 630 was sold new in Lindsborg, Kansas, and has never strayed more than 10 miles from the dealership. It, too, was taken down to the bare bones before being meticulously restored. Another plus is that it's loaded with most of the popular options.

## Model 730

When the Model 730s were rolled out, they were mechanical twins to the 720s but wore a new suit of duds—new tinwork with slanted dash and steering wheel, just like their siblings. These were the last tractors that sprang from the Model G lineage: Models G, GM, 70, 720, and finally the Model 730, Deere's biggest row-crop tractor to date.

Vehicle variants included Row-Crop, Standard, and Hi-Crop, with all versions available with gasoline, all-fuel, LP-gas, or diesel powerplants.

Standard equipment and optional equipment were similar to the Model 720s. Like the Model 720, the 730 was offered through the Industrial Division and could be had in industrial yellow, John Deere yellow, agricultural green and yellow, or any color the customer requested as long as he paid the up-charge for special paint. However, an Industrial version per se didn't happen.

There were a couple of interesting deviations about the Model 730. Production of the Model 730 tractor at the Waterloo, Iowa, plant continued longer than other 30 Series tractors. In the case of the 730, it rolled down the production line until March 1, 1961. However, all tractors built after February 1960 were exported.

The 730 wasn't done yet. Deere & Company's new facility in Rosario, Argentina, began producing (assembling) the Model 730 in 1959. Most components were shipped from Waterloo, but some were provided by local manufacturers. At the close of Model 730 production in Waterloo, the assembly line and production tooling were shipped to the factory at Rosario. Once installed, full production began on all variants of the Model 730—Row-Crop, Standard, and Hi-Crop. Production continued until sometime in 1970, with 20,000-plus Argentina Model 730s manufactured. There's still more. Additional Model 730s were assembled at Deere & Company's Monterrey, Mexico, plant. The Model 730 was available with steel front and rear wheels via special order.

Look under the Model 720 for Nebraska test results. Test numbers are 593, 594, 605, and 606.

## Model 830 Diesel

Follow the lineage of the Model 830, and you'll eventually arrive back at the Waterloo Boy, which was Deere's first two-cylinder tractor. In between are Models D, R, 80, and 820—all big-boy tractors designed primarily for one purpose: tough drawbar and belt work. Mechanically, it was a duplicate of the late, or black-dash, Model 820.

Standard was standard, period. The only other Model 830 vehicle was the Industrial version. Until early 1959, the Model 830 could be ordered through the Industrial Division with heavy-duty front- and rear-axle packages, thus, it became an Industrial model. The rest of the options were the same as the agricultural tractors.

In January 1959, the Industrial version became official. The Model 830-I came with the heavy-duty axle package as standard equipment, along with industrial yellow paint. Post-January 1959 production of the 830-I tallied about 127 copies.

Adjustable front axle, three-point hitch, styled rear fenders with lights, Float Ride seat, pre-cleaner and air stack, dish-type rear wheel, dual hydraulics, power steering, and V-4 starting engine are some of the "desirables" on this 1960 Model 730 Diesel.

Standing face-to-face with the big guy. This 1959 Model 830 Diesel was the ultimate two-cylinder John Deere. Outfitted with Power Steering, PTO, Powr-Trol, and front- and rear-wheel weights, the Model 830 tipped the scales at about 9,875 pounds. Although the Model 830 wasn't tested at Nebraska, in test number 632, conducted on a Model 820, they managed to hang enough iron on the tractor to boost the weight to an impressive 11,995 pounds.

The owner stated that this Model 830 was one of his earlier restorations, which he did back in 1988. No apologizes needed since it looks showroom slick 15 years later.

Like the other 30-Series Waterloo tractors, the Model 830 was a basic, bare-bones, 2x4 tractor—two-cylinders and four wheels. The buyer could custom design his dream tractor from the long options list: Power steering, Float-Ride seat, All-Weather steel cab, Weather Brake or Weather-Brake cab, foot-operated throttle, cigarette lighter, custom Powr-Trol, live PTO, and front- or rear-wheel weights.

Another decision for the buyer was to decide if he wanted his Model 830 tractor equipped with pony engine start or electrical start. Front and rear steel wheels could be special-ordered for the Model 830.

The Model 830 was never tested at Nebraska, but the test lab did verify that because the 830 tractor carried the same engine and drivetrain as the 820 model, it produced the same 67 drawbar and 72 belt horsepower.

## Model 840

Another two-cylinder Deere tractor is the Model 840. This tractor was a special Industrial vehicle converted from the Model 830. Hancock Manufacturing Company, in Lubbock, Texas, modified the Model 830 to power a small elevating scraper for light, fast, industrial, or agricultural dirt-moving applications.

The operator's platform was moved forward of the rear axle and offset to the left. This allowed the scraper's gooseneck to attach directly over the tractor's rear axle.

The Hancock firm made perhaps 60 to 70 units before Deere took the conversion process in-house.

Most of these were built in 1959, 1960, and 1961, although it isn't known just when production ended. Some sources say 1964. Probably somewhere between 800 and 1,200 units were manufactured.

This 1959 Model 435 Diesel in this setting could move a collector to unrestrained tears—or envy.

### Model 435

Introduced in 1959, the Model 435 has the distinction of being the last agricultural tractor to join the John Deere two-cylinder line. Some might argue that it really isn't a true John Deere, because the exhaust note sounds very different and the engine isn't a John Deere product.

However, except for the engine, the Model 435 is essentially a Model 430 Row-Crop Utility with different footwear. The engine is a vertical two-cylinder, which happens to be a GM-made two-cycle diesel.

The supercharged engine has a 3.875x4.5 -inch bore and stroke, which amounts to 106 cubic inches of displacement with a rated rpm of 1,850. The standard transmission was a 4/1, with a 5/1 offered as an option. Total production was 4,626 copies from March 31, 1959, to February 29, 1960.

Deere's first small diesel-engined tractor was tested at Nebraska in 1959 and generated 28 drawbar and 32 PTO horsepower, according to test number 716.

### Two-Cylinder Deere Denouement

The two-cylinders carried the load for Deere & Company tractor program for just over four decades. By the end of the 1950s, Deere's competition considered the two-cylinder tractors, especially the 30 Series,

This is one of 4,626 Model 435 Diesels produced at Deere & Company's Dubuque, Iowa, facility. Yes, the silver rims are correct. All Dubuque tractors equipped with the Power Adjust rear wheels were sent out the door with silver rims.

outdated and obsolete. However, as planned, the new look of the 30 Series convinced the majority of dealers, farmers, and competitors that the two-cylinder was going to be around for many more years. They were right, because today, John Deere 30 Series tractors are some of the most sought-after tractors on the planet. In fact, as far as collectors are concerned, most would probably agree that there's not a bad tractor in the entire John Deere two-cylinder lineup, from the Waterloo Boy to the 30 Series.

John Deere is as much a part of the United States' history as Meriwether Lewis and William Clark, Davy Crockett and Jim Bowie, or Wilbur and Orville Wright.

If you have a two-cylinder John Deere tractor, you own a piece of U.S. history just as authentic as Colt and Winchester, Ford and Chevy, or Maytag and Frigidaire. They're all keepers.

# Bibliography


Arnold, Dave. *Vintage John Deere.* Stillwater, Minnesota: Voyageur Press, 1995.

Baldwin, Nick. *Classic Tractors of the World.* Stillwater, Minnesota: Voyageur Press, 1998.

Beemer, Rod and Chester Peterson Jr. *Inside John Deere: A Factory History.* Osceola, Wisconsin: MBI Publishing Company, 1999.

Bowman, John S., ed. *The American West Year by Year.* New York, New York: Crescent Books, 1995.

Boyer, Paul S., ed. *The Oxford Companion to United States History.* New York, New York: Oxford University Press, 2001.

Broehl, Wayne G. *John Deere's Company: A History of Deere & Company and Its Times.* New York, New York: Doubleday & Company, 1984.

Brown, Theo. *Deere & Company's Early Tractor Development.* Grundy Center, Iowa: Two-Cylinder Club, 1997.

Deere & Company. *How Johnny Popper Replaced the Horse.* Moline, Illinois: Deere & Company, 1988.

Deere & Company. *John Deere Tractors: 1918-1994.* St. Joseph, Michigan: ASAE, 1994.

Dunning, Lorry. *John Deere Tractor Data Book.* Osceola, Wisconsin: MBI Publishing Company, 1996.

Dunning, Lorry. *Ultimate American Farm Tractor Data Book.* Osceola, Wisconsin: MBI Publishing Company, 1999.

Gray, R. B. *The Agricultural Tractor: 1855–1950,* St. Joseph, Michigan: ASAE, 1954.

Hobbs, J. R. *John Deere Tractors: First Numbered Series.* Bee, Nebraska: Green Magazine, 1996.

Hobbs, J. R. *The John Deere 20 Series.* Bee, Nebraska: Green Magazine, 1996.

Hobbs, J. R. *The John Deere Styled "Letter" Series, 1939–1952.* Bee, Nebraska: Green Magazine, 1997.

Kirshon, John W., ed. *Chronicle of America.* Mount Kisco, New York: Chronicle Publications, 1989.

Larsen, Lester. *Farm Tractors 1950–1975.* St. Joseph, Michigan: ASAE, 1981.

Leffingwell, Randy. *Classic John Deere Tractors.* Osceola, Wisconsin: MBI Publishing Company, 1994.

Leffingwell, Randy. *John Deere Farm Tractors.* Osceola, Wisconsin: MBI Publishing Company, 1993.

Leffingwell, Randy. *The American Farm Tractor.* Osceola, Wisconsin: MBI Publishing Company, 1991.

Macmillan, Don and Roy Harrington. *John Deere Tractors and Equipment, 1960–1990.* Vol. 2. St. Joseph, Michigan: ASAE, 1991.

Macmillan, Don and Russell Jones. *John Deere Tractors and Equipment, 1837–1959.* Vol. 1. St. Joseph, Michigan: ASAE, 1988.

Macmillan, Don. *John Deere Tractors Worldwide: A Century of Progress 1893–1993.* St. Joseph, Michigan: ASAE, 1994.

Macmillan, Don. *The Big Book of John Deere Tractors.* Stillwater, Minnesota: Voyageur Press, 1999.

Miller, Marilyn and Marian Faux, eds. *The New York Public Library American History Desk Reference.* New York, New York: Stonesong Press, 1997.

Pripps, Robert N. *Big Green: John Deere GP Tractors.* Osceola, Wisconsin: MBI Publishing Company, 1994.

Pripps, Robert N. *Illustrated Buyer's Guide: John Deere Two-Cylinder Tractors.* Osceola, Wisconsin: MBI Publishing Company, 1992.

Pripps, Robert N. *John Deere Photographic History.* Osceola, Wisconsin: MBI Publishing Company, 1995.

Rukes, Brian and Andy Kraushaar. *Original John Deere Model A.* Osceola, Wisconsin: MBI Publishing Company, 2000.

Sanders, Ralph W. *Ultimate John Deere. Stillwater, Minnesota*: Voyageur Press, 2001.

Wendel, C. H. *Nebraska Tractor Tests Since 1920.* Osceola, Wisconsin: MBI Publishing, 1993.

# Index

## Other MBI Publishing Company titles of interest:

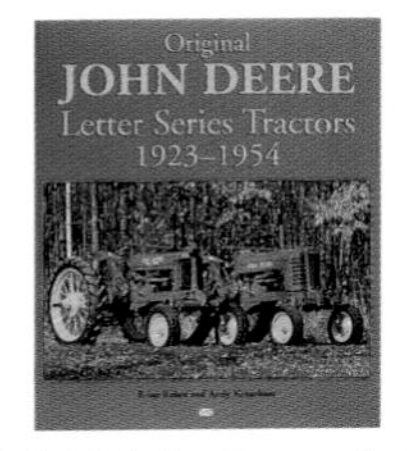

**Original John Deere Letter Series Tractors 1923–1954**
ISBN 0-7603-0912-4

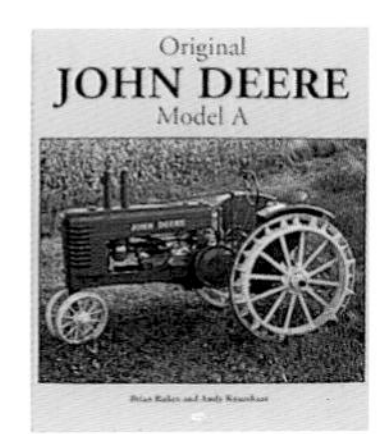

**Original John Deere Model A**
ISBN 0-7603-0218-9

**John Deere: The Classic American Tractor**
ISBN 0-7603-1365-2
153108

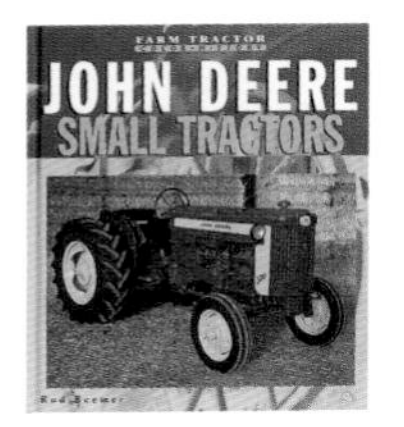

**John Deere Small Tractors**
ISBN 0-7603-1130-7
134869

**John Deere New Generation Tractors**
ISBN 0-7603-0427-0

**John Deere Industrials**
ISBN 0-7603-1023-8

**John Deere Two-Cylinder Collectibles**
ISBN 0-7603-0750-4

**John Deere Tractor Data Book**
ISBN 0-7603-0228-6

**John Deere Photographic History**
ISBN 0-7603-0058-5

**Find us on the internet at www.motorbooks.com 1-800-826-6600**